Ella Rodman Church

Some useful Animals

Ella Rodman Church

Some useful Animals

ISBN/EAN: 9783337240592

Printed in Europe, USA, Canada, Australia, Japan

Cover: Foto ©berggeist007 / pixelio.de

More available books at **www.hansebooks.com**

W. C. Morys.

AN EASTERN CARAVAN.

Frontispiece.

SOME USEFUL ANIMALS.

BY

ELLA RODMAN CHURCH,

AUTHOR OF "BIRDS AND THEIR WAYS," "FLYERS AND CRAWLERS,"
"FLOWER-TALKS AT ELMRIDGE," ETC.

———•———

PHILADELPHIA :
PRESBYTERIAN BOARD OF PUBLICATION
AND SABBATH-SCHOOL WORK,
1334 CHESTNUT STREET.

WESTCOTT & THOMSON,
Stereotypers and Electrotypers, Philada.

PREFACE.

THE "Elmridge Series" of books is proving very popular. Young people are learning many interesting facts about the creatures God has made. The books make no pretensions to exhaustiveness as scientific treatises or manuals, but aim only to give such information as all intelligent young persons should have concerning the various objects considered, and to give it in a pleasant way that makes the task of reading an easy one.

The present volume treats thus conversationally of "some useful animals." The children and young folks in many other homes have thus the privilege of enjoying in these bright pages what the little people of the Kyle home first enjoyed with their delightful young governess.

J. R. M.

Philadelphia, 1888.

CONTENTS.

8 *CONTENTS.*

ILLUSTRATIONS.

10 *ILLUSTRATIONS.*

SOME USEFUL ANIMALS.

CHAPTER I.

LONG-EARED FRIENDS.

IT was a rainy day at Elmridge, and the children were deeply interested in some new books which Miss Harson had wisely kept for just such an occasion as this. There were plenty of pictures, and every little while there would be an exclamation of delight from one or other of the readers, with an appeal to their governess as to the exact meaning of certain things.

Miss Harson had a book too, and was supposed to be reading; but she presently decided that it would be as well to lay aside her own volume and devote herself to her pupils.

"I do like pictures of donkeys," Edith

was saying; "they look so nice and patient. And I wish—"

"That you were a donkey yourself?" asked Malcolm, mischievously.

"No," replied his little sister, feeling rather hurt at the question; "you know I did not mean that. But I almost wish I was a little English girl, because then I would have a donkey to ride on."

"Then you wouldn't have Miss Harson," said Clara, very soberly, "and you wouldn't have me, and—and all of us."

"Oh, I don't want to, really," exclaimed Edith, in great anxiety lest they should all have thought her in earnest; "I wouldn't go away from Elmridge for the world. But it's nice to have a donkey to ride on; it looks like a queer little horse.—Doesn't it, Miss Harson?"

"And a queer little horse you'd find it, dear," was the laughing reply, "although it seems so attractive in this pretty English story. It is a species of horse, though, and I think we might all enjoy finding out something about it. Are you ready for more animals?"

The little Kyles were always ready for Miss Harson's delightful " talks," and, having spent more than an hour over their books, they had become pretty well acquainted with their contents; so there was a general drawing up around the young lady with an evident expectation of good things to come.

"Isn't the donkey some relation to the horse?" asked Malcolm.

"He is a first cousin," was the reply, "but these relatives do not agree very well. The horse looks down upon the humbler donkey, or ass, who is certainly not handsome with his big head and ears and his small, insignificant body. His tail, too, is very different from that of the horse, as in the latter animal it is a very graceful appendage—a sweeping plume of hair; but the donkey's tail has very little hair, and that only at the end. Instead of having the horse's long, flowing mane, which hangs down on one side of its neck, the poor donkey shows only a ridge of short, stubby hair."

Yes, they could see it all in the picture,

and it was not much to be wondered at that the horse should not care to claim such a shabby-looking relative.

"He isn't very pretty, I suppose," said Edith, "but he looks good."

"Perhaps, dear, you will not think him

THE DONKEY.

'good' when you hear that he bites and kicks when angry, and is considered a stubborn creature."

This was certainly not after the Elmridge standard of goodness, and the little girl looked quite disappointed at the bad character of the donkey.

"It is scarcely fair," continued her governess, "to tell you such things without also mentioning the good traits which the animal possesses. It is said of him that, 'though his nature is stubborn, he has many good qualities. He is gentle and patient; he is fond of his master when his master is kind to him; I think he seems much cleaner than most animals, for he will not drink water if it is dirty, and, however much he is neglected, he never has vermin; he hates wetting his feet, and even when loaded will go round to get away from the dirty parts of the road.'"

"But, Miss Harson," said Malcolm, "isn't the donkey very stupid?"

"Not always, although the expression 'a perfect donkey' is sometimes applied to a silly person. The animal is said to be really more intelligent than the horse, and an English farmer who had several horses and one donkey said that whenever these animals played him an ingenious trick the donkey was sure to be the ringleader. This was shown very plainly once, when the farmer fastened up several of his horses with the donkey in a large field next to one in which there was a fine crop of oats nearly ripe. The farmer found to his surprise that the prisoners had contrived to get among the oats, of which they ate a large quantity and trampled much more, but how they managed it he could not see. Keeping watch early one morning, he was rewarded by the unexpected sight of the donkey deliberately undoing the fastenings of the gate with his teeth, as though he had always been accustomed to do it, and letting his companions out to breakfast on oats."

" I'm sure that wasn't stupid," said Clara.

" No, indeed!   And some of these much-abused animals, it seems, have even been taught to perform tricks in public.   Fancy such a great creature, for instance, drinking out of a glass !"

The children could not imagine how a donkey could possibly drink out of anything so small; but when Miss Harson read them the following account out of a queer old book, they were still more surprised:

" 'There was a cunning player in Africa, in a city called Alcair, who taught an asse divers strange tricks or feats, for in a publick spectacle, turning to his asse (being on

A NEW GATEKEEPER.

a scaffold to shew sport), said, "The great
sultan proposeth to build him an house, and
shall need all the asses of Alcair to fetch
and carry wood, stones, lime, and other
necessaries for that business." Presently,
the asse falleth down, turneth up his heels
in the air, groaneth and shutteth his eyes
fast, as if he had been dead. While he lay
thus the player desired the beholders to
consider his estate, for his asse was dead.
He was a poor man, and therefore moved
them to give him money to buy another asse.

"'In the mean time, having gotten as much
money as he could, he told the people that
he was not dead, but, knowing his master's
poverty, counterfeited in that manner, whereby he might get money to buy him provender; and, therefore he turned again to his
asse and bid him arise, but he stirred not at
all. Then did he strike and beat him sore
(as it seemed) to make him arise, but all in
vain: the asse laid still. Then said the
player again, "Our sultan hath commanded
that to-morrow there be a great triumph
without the city, and that all the noble
women shall ride thither upon the fairest

asses, and this night they must be fed with
oates, and have the best water of Nilus to
drink." At the hearing whereof, up started
the asse, snorting and leaping for joy.'

"What do you think of that?" asked their
governess.

The audience were quite wild to see such
a wonderful ass, but Miss Harson told them
laughingly that she had not the least idea
where they would go to find one.

"We must not call the animal stupid, I
think," she continued, "as he is only so
under bad treatment. One of our favorite
naturalists says, 'Let any one turn an old
ass into a field and try to mount and ride
him, and after an hour or so the ass will not
appear a very stupid animal. It is on such
an occasion as this that kicking and biting
are indulged in, while the animal jumps
about in all directions to prevent his being
mounted. Should an adventurous rider
succeed in getting on his back, Master
Donkey stands perfectly still, or else he
wriggles and shakes his burden off upon
the ground, and then runs away. If this
cannot be accomplished and man or boy

persists in sticking, the animal will lie down and roll over. Another pleasant habit with an unwelcome burden is to grind his leg against a wall or the rough stem of a tree, also to walk into a pool and lie down there.'"

"Then what does the man or boy do?" asked Clara, eagerly.

"Gets out the best way he can, I should think," was the reply, "and with a better opinion of the intelligence of donkeys than he had before. A story is told of a donkey at Carisbrooke Castle, in the Isle of Wight, who used to draw water by a large wheel from a deep well. When his owner would say, 'Tom, my boy, I want water; get into the wheel, my good lad!' the animal immediately obeyed, and seemed to know just how many times the wheel should turn upon its axis to bring the bucket up, for every time it reached the top of the well he stopped. He would then turn his head round to see when his master laid hold of the bucket to draw it toward him, as he had then either to go back or to come forward a little. He never made a mistake."

Of course this was "a dear old donkey," and there was a general feeling among the little Kyles that such an animal would be just about the nicest and most useful pet that could be found.

"But we haven't got any wheel to turn," suggested Malcolm.

"Never mind," replied his sisters; "he could do something else, then."

"But not all donkeys are so capable," said Miss Harson: "many of them are both stupid and obstinate; and if you really owned one, you would probably get out of patience with him every hour in the day. Only those who can properly govern themselves are fit to train animals, and harsh treatment makes a donkey especially vicious. It is said, when young, to be sprightly, and even pretty, but it soon gets slow and stupid. Yet it seems to like its owner when not cruel, as is too often the case; it can scent him at a distance, tell him at once from others, and seems to know just where he has passed and the places at which he stops. 'When overloaded, it manifests its sense of injury by hanging down its

head and flapping its ears; and when hard
pressed, it opens its mouth and draws back
its lips with a ghastly grin. If blinded, it
will remain motionless, however easy it
might be to remove the impediments that
hinder its sight. It walks, trots and gallops
like a horse, but, though it sets out freely,
it is soon tired, and requires to be managed
with some address to make it proceed.'"

"Don't donkeys often draw carts?" asked
Malcolm. "In one of my books is a funny
picture of one with a little dog on his back.
Here it is. The cart seems to be full of
vegetables and things, and the picture is
called 'Costermonger's Donkey and Dog.'"

"What a queer word!" exclaimed Clara
as she repeated "costermonger." "What
does it mean, Miss Harson?"

"Very much what 'huckster' does with
us," replied the young lady.—"And the
'things,' Malcolm, of which you spoke, appear to be fruit of various kinds. I think I
can distinguish apples, plums, and goose-
berries probably, for these berries are
quite plentiful in England, and this is an
English picture."

"But what is the dog doing on his back?" asked Edith. "How cunning he looks there! And the donkey seems to be turning around to speak to him."

"Here is the little story, dear," said Miss Harson: "'A costermonger who was kind-hearted to both his donkey and his dog had trained them so well that whenever he had to leave his cart and carry his vegetables inside a house the dog instantly mounted guard by jumping on Jack's back. The donkey never offered to stir until his master appeared; but when he came in sight, the dog jumped down, and the donkey started for the next customer's house.'"

This was quite as delightful as the wheel-story, and the children were beginning to feel quite well acquainted with donkeys.

"This animal," continued their governess, "has been called in England 'the poor man's horse' because it costs so little to feed or shelter him, and he can be made useful in many ways. Carrying vegetables to market is his common occupation, and a bunch of thistles or a few handfuls of grass will satisfy his hunger. A little kind treatment goes a great way with him, and he knows when he is being sympathized with as well as any of us.   Here are some verses which describe a donkey with a kind-hearted young master:

"'OLD JACK THE DONKEY.

"'Old Jack was as sleek and well-looking an ass
As ever on common munched thistle or grass,
And, though 'twas not gaudy, that jacket of brown
Was the pet of the young and the pride of the town.

"'And, indeed, he might well look so comely and trim,
When his young Master Joe was so gentle to him;
For never did child more affection beget
Than was felt by young Joe for his four-footed pet.

"'Joe groomed him and fed him, and each market-day
Would talk to his darling the whole of the way,
And Jack before dawn would be pushing the door,
As though he would say, "Up, Joe! slumber no more."

"'One day Jack was wandering along the roadside,
When an urchin the donkey maliciously eyed,
And, aiming too surely at Jack a sharp stone,
It struck the poor beast just below the shin-bone.

"'Joe soothed and caressed him and coaxed him, until
They came to a stream by the side of a hill,
And with the cool water he washed the swelled limb,
And after this fashion kept talking to him:

"'"Poor Jack! did he pelt him, the coward so sly?
I wish I'd been there with my stick standing by!
It does not bleed now; 'twill be well in a trice.
There! let me just wash it. Now isn't it nice?"

THE SWELLED LIMB.

" ' And Jack nestled down with his soft velvet nose,
As soon as he could, under Joe's ragged clothes,
And he looked at his master as though he would say,
" I'm sure I can never your kindness repay." ' "

These verses were much liked, but little
Edith appeared to think that she would
scarcely like to have a donkey as close to

her as if it were a dog; she was afraid it might kick.

"It only kicks," was the reply, "when it does not want people to catch it or to get on its back; but it is said to be very affectionate and easily managed by kindness. It is nearly always a hard-worked animal, and seems to get little time for eating or sleeping. In some parts of Ireland these little drudges are made to draw carts from mines and back again. When the load is emptied, the driver just starts the cart in the right direction, and then lies down and goes comfortably to sleep, leaving the wise donkey to find his way back by his own sagacity. Here is a little story of some children who made good use of a donkey and a donkey-cart."

The story was called

THE HAPPY BIRDS.

The Bird children were a happy little flock. They lived in a little house with only four rooms in it, but it was warm and neat, and bright with father- and mother-love, which is always like sunshine in a home. Father and Mother Bird had to work hard, it is true, but they loved to work, and the children loved to help. Richard, the eldest, was able to help a good deal, and the next boy, Christopher, could help a little; and, as for Margie, I do not know how mother could have got along without her; and dear little Rosie—why, she was as important as any of them: they could not get along, of course, without her, for babies help—do you not know they do?—in a home.

These children did not often have a ride: Donkey Bray was almost always in use for some work about the farm. I do not know but he had to work harder than any of the rest, yet he was well fed and loved; and when he would not go, they did not whip him. No, no, no! They gave him some oats and coaxed him to go, and after a little he would begin again quite bravely. Mr. Bird did not believe in whipping child or beast; he had a way of loving them into doing what he wanted.

This afternoon father was going away with the minister to see a sick man. Be-

fore he went he harnessed Donkey Bray
into the wagon and told the children they
might have a ride.

"Where shall we go?" shouted Chris.
"Let's ride over to see the Smiths."

"No," cried Margie; "let's go down by
the pond and get flowers. I know there are
some there by this time, for it's two weeks
since the snow melted away, and I'm sure
the violets have had plenty of time to
grow."

"Now, let me tell you," said Richard,
"mother's been baking to-day; and let's
take a loaf of bread and a pie over to poor
sick Hannah Dean."

"Why, that's 'most to the pond," said
Margie.

"So it is, and we can go that way and
get some flowers and carry her some," said
Chris.

"All right," said Richard; "so we will."

Mother very willingly gave them a loaf
and a pie, and added a cup of cranberry
jelly; so off they started. Donkey did not
step off very rapidly, but that gave them
the more time to enjoy the sweet spring air

THE HAPPY BIRDS.

and to chatter and sing and laugh. Donkey stepped by degrees over the two miles to the pond, where they found a few violets and a dear little blue flower that grows under the old leaves heaped up around the roots of trees, and then they drove over to Hannah's.

Poor Hannah was a cripple. She had had the rheumatism a long time, and her hands and feet and limbs were all drawn up; so that she lay in a heap in the bed and could not straighten herself out. Hannah was delighted, of course, with the children and with what they had brought. She was feeling pretty well to-day and had a nice chat with them, and kissed and patted the baby; and when it was time for them to go, she said,

"Now I want you all to stand in a row right there at the foot of my bed and sing."

They knew what to sing. "And let this feeble body fail," was a favorite hymn of hers, and the Birds had learned it on purpose to sing to her.

"I like that last verse best of all," she

said. "Sing it over once more, darlings, if you're not too tired."

The children were never tired of singing to Hannah; so they repeated the verse, and then they bade her "Good-bye" and got into the wagon and drove home.

The children thought this was a very nice little story, and Edith wished she could go in a donkey-cart with some jelly and flowers to see a sick old woman.

"Miss Harson," said Malcolm, "do people ever ride on donkeys?"

"Not in our country, and not very often in England except among children. They appear to be fond of these queer little steeds, and look forward to donkey-riding as one of the great pleasures of going to the seashore. An entertaining writer says that sweeps and dust-boys in England perch themselves on the end of the animal's spine and seem highly delighted with this uncomfortable seat. They seem to find much amusement in thus riding.

"It is in warm countries that the finest of these animals are found, as cold weather

does not agree with them, and they are seldom seen in the United States.   In Spain, for
instance, which has been called the beginning of Arabia and Africa, the donkey is
so much larger, stronger and handsomer
that he is almost equal to the horse.   The
flower-sellers, who are quite an important
class, use them to carry their merchandise
in panniers, or deep baskets, and they are
fond of decorating their heads with tassels
and other gay trappings.   These asses are
so well fed and cared for that they are
really fine-looking animals, and nothing is
heard of their being stupid and obstinate.
As to that, here in one of your books is
this anecdote: 'A boy who sold vegetables
in London used in his employment an ass
which conveyed his baskets from door to
door.   Frequently he gave the poor industrious creature a handful of hay or greens
by way of reward.   The boy had no need
of any goad for the animal, and seldom indeed had he to lift up his hand to drive it
on.   This kind treatment was one day re-
ferred to by some one, and he was asked
whether his beast was stubborn.   "Ah,

THE YOUNG VEGETABLE-SELLER.

master!" he replied; "it's of no use to be
cruel, and, as for his stubbornness, I cannot
complain, for he is ready to do anything
or go anywhere. I bred him myself. He
is sometimes playful, and once ran away
from me. You will hardly believe it, but
there were more than fifty people after
him, attempting in vain to stop him; yet,
after all, he turned back of himself, and
never stopped till he ran his head into
my bosom.'''"

"Why don't people always treat donkeys
kindly, then?" asked Clara. "They always
seem to be good with good masters."

"Ah, my dear child!" replied her governess; "why do not people always treat
people kindly? They certainly behave better under such treatment, and even de-
praved criminals have been reclaimed by
love. But how few act on this plan! Let
us try to remember that kindness is never
wasted, but brings forth fruit in the most
barren soil: 'A righteous man regardeth
the life of his beast: but the tender mercies
of the wicked are cruel.'

"Besides the Spanish donkeys," contin-

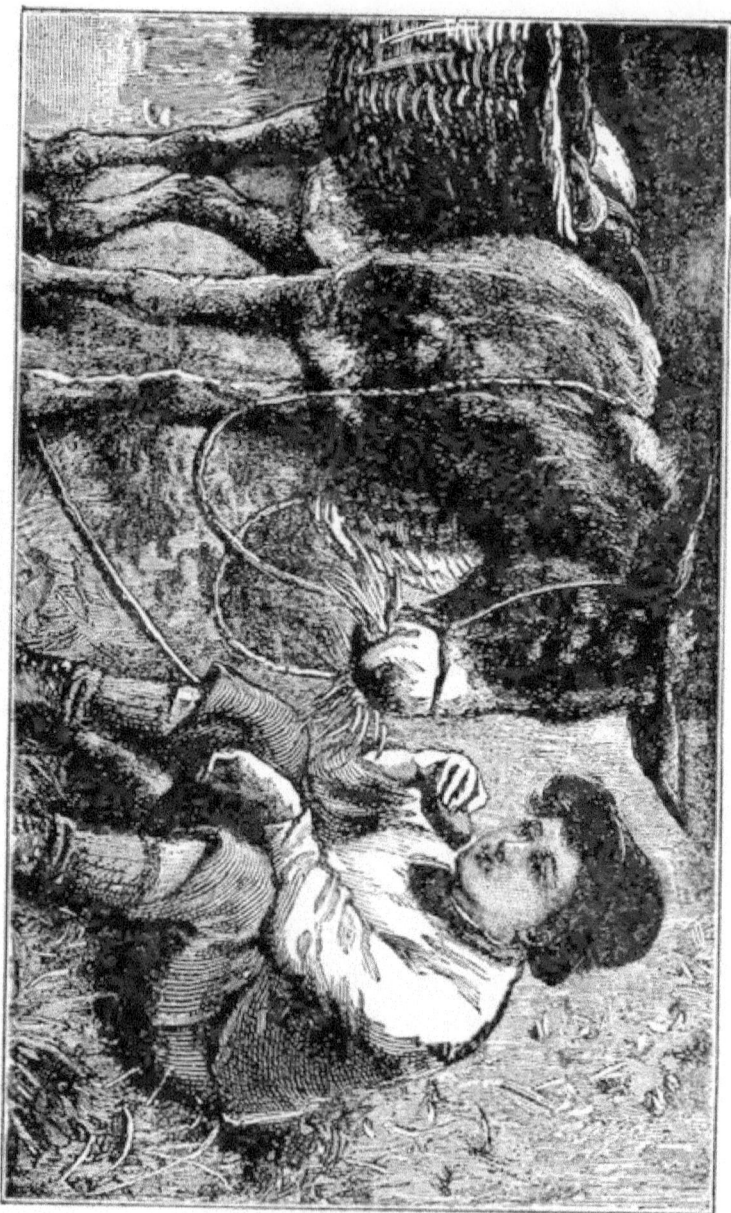

KIND TREATMENT.

ued the young lady, "Arabian and Egyptian donkeys are very handsome animals, and in Arabia and Egypt they are used in place of horses by the rich and the great. The Arabian asses are considered the finest, as they carry their heads gracefully and throw out their legs in quite an elegant fashion in walking or galloping. They are beautifully decorated, too, and their 'housings' are very showy. Travelers always have a great deal to say about these Eastern donkeys, and I have here a description which is rather long and meant for grown people. Do you think you will care to hear it?"

" Rather long and meant for grown people"! Why, that was exactly what the children wanted, and, trying to look as much like grown people as possible, the little Kyles settled themselves for enjoyment.

"This is written of Cairo, in Egypt, where 'donkey-riding is universal, and no one thinks of going beyond the Frank'—or Christian—'quarters on foot. If he does, he must submit to be followed by no less than six donkeys with their drivers. A

friend of mine was attended by such a
cavalcade for two hours, was obliged to
yield at last, and made no second attempt.
When we first appeared at the gateway of
a hotel equipped for an excursion, the rush
of men and animals was so great that we
were forced to retreat until our servant
and the porter whipped us a path through
the yelling and braying mob. After one or
two trials I found an intelligent Arab boy
named Kish who for five piastres a day
furnished strong and ambitious donkeys,
which he kept ready at the door from
morning to night. The other drivers re-
spected Kish's privilege, and thenceforth I
had no trouble.

"'The donkeys are so small that my feet
nearly touched the ground, but there is no
end to their strength and endurance. Their
gait, whether in pace or in gallop, is so easy
and light that fatigue is impossible. The
drivers take great pride in having high-
cushioned red saddles, and in hanging bits
of jingling brass to the bridles. They keep
their donkeys close-shorn, and frequently
beautify them by painting them various

colors. The first animal I rode had legs
barred like a zebra's, and my friend's rejoiced in purple flanks and a yellow belly.
The drivers run behind them with a short
stick, punching them from time to time or
giving them a sharp pinch on the rump.
Very few of them own their donkeys,
and I understood their pertinacity when I
learned that they frequently received a
beating on returning home empty-handed.

"'The passage of the bazaars seems at
first quite as hazardous on donkey-back as
on foot, but it is the difference between
knocking somebody down and being knocked down yourself; and one certainly prefers
the former alternative. There is no use in
attempting to guide the donkey, for he will
not be guided. The driver shouts behind,
and you are dashed at full speed into a confusion of other donkeys, camels, horses,
carts, water carriers and footmen. In vain
you cry out " Bess " (" Enough)," " Piacco,"
and other desperate abjurations : the driver's
reply is, " Let the bridle hang loose !" You
dodge your head under a camel-load of
planks; your leg brushes the wheel of a

dust-cart; you strike a fat Turk plump in
the back; you miraculously escape upsetting a fruit-stand; you scatter a company
of spectral, white-masked women, and at
last reach some more quiet street with the
sensations of a man who has stormed a
battery.

"'At first this sort of riding made me
very nervous, but presently I let the donkey go his own way, and took a curious
interest in seeing how near a chance I ran
of striking or being struck. Sometimes
there seemed no hope of avoiding a violent
collision, but by a series of most remarkable
dodges he generally carried me through in
safety. The cries of the driver behind me
gave me no little amusement: "The howadji
comes! Take care on the right hand! Take
care on the left hand! O man, take care!
O maiden, take care! O boy, get out of
the way! The howadji comes!" Kish had
strong lungs, and his donkey would let
nothing pass him; and so, wherever we
went, we contributed our full share to the
universal noise and confusion.'"

This seemed very funny to the children;

and when Miss Harson had finished reading, they asked her what "howadji" means.

"It is the same as 'lord,' 'master' or
'gentleman,'" was the reply, "and it sounds
oddly enough to those who have not traveled in the East. But we shall have to stop
traveling for this afternoon, for here comes
Jane to say that dinner is ready."

This was quite a surprise, for they had all
been too much interested to feel hungry.

CHAPTER II.

"I HOPE, Miss Harson," said Edith, in a very coaxing way, "that you are going to tell us some more about these nice donkeys?"

"I shall want you to tell me something about them first," was the smiling reply. "Who can remember where the donkey, or ass, is mentioned in the Bible?"

"I remember Balaam's ass," said Clara, "but I can't remember just where it is spoken of."

"Then I must help you a little. You will find the account, I think, in the twenty-second chapter of Numbers."

Having found the place, Clara read as Miss Harson requested her, from the twenty-first to the thirty-fourth verses:

"'And Balaam rose up in the morning, and saddled his ass, and went with the

41

princes of Moab. And God's anger was kindled because he went; and the angel of the Lord stood in the way for an adversary against him. Now he was riding upon his ass, and his two servants were with him. And the ass saw the angel of the Lord standing in the way, and his sword drawn in his hand; and the ass turned aside out of the way, and went into the field; and Balaam smote the ass, to turn her into the way. But the angel of the Lord stood in a path of the vineyards, a wall being on this side, and a wall on that side. And when the ass saw the angel of the Lord, she thrust herself unto the wall, and crushed Balaam's foot against the wall; and he smote her again. And the angel of the Lord went further, and stood in a narrow place, where there was no way to turn, either to the right hand or to the left. And when the ass saw the angel of the Lord, she fell down under Balaam: and Balaam's anger was kindled, and he smote the ass with a staff. And the Lord opened the mouth of the ass, and she said unto Balaam, What have I done unto thee, that thou hast smitten me these three times? And Balaam said unto the ass, Because thou hast mocked me: I would there were a sword in mine hand, for now would I kill thee. And the ass said unto Balaam, Am not I thine ass, upon which thou hast ridden ever since I was thine unto this day? was I ever wont to do so unto thee? And he said, Nay. Then the Lord opened the eyes of Balaam, and he saw the angel of the Lord standing in the way, and his sword drawn in his hand, and he bowed down his head, and fell flat on his face. And the angel of the Lord said unto him, Wherefore hast thou smitten thine ass these three times? behold, I went out to withstand thee, because thy way is perverse before me: and the ass saw me, and turned from me these three times; unless she had turned from me, surely now also I had slain thee, and saved her alive.' "

"How mean Balaam must have felt," said Malcolm, "when he found out that he had been whipping the ass for saving his life!"

"But he must have been awfully frightened," added Clara, "to hear her speak.— It sounds like a fairy-tale, Miss Harson."

"It is the only place in the Bible," said their governess, "where an animal is represented as speaking, but we must remember this incident when we are disposed to regard the ass as a particularly stupid animal. A man who saw the angel of the Lord could not well be frightened by anything else that might happen, and this, I think, accounts for Balaam's taking the speaking of his ass as though it had been an every-day affair. The ass is often mentioned in the Bible, for in Eastern countries it is used as we use a saddle-horse by people of the highest rank, who decorate it with a very handsome saddle and harness. If you will find the twenty-second chapter of Genesis, Malcolm, and read the third verse, you will see that Abraham, who was a rich chieftain and prince, used this animal for riding."

Malcolm read:

"'And Abraham rose up early in the morning, and saddled his ass, and took two of his young men with him, and Isaac his son, and clave the wood for the burnt offering, and rose up and went unto the place of which God had told him.'"

"This was that melancholy journey to offer up 'his son, his only son Isaac,' which had so happy an ending. In Judges, tenth chapter and fourth verse, it is said of Jair the Gileadite that 'he had thirty sons that rode on thirty ass-colts, and they had thirty cities.' Another judge had 'forty sons and thirty nephews that rode on threescore and ten ass-colts;' which shows plainly that these animals were used by the rich and the great in place of horses."

"But they had horses," said Clara, "because you told us, Miss Harson, of so many places in the Bible where horses are mentioned."

"Yes, dear, that is very true; but the horse was especially a war-animal, as has also been shown, while the ass was used more for journeys and purposes of peace. Our blessed Lord on his entry into Jerusalem, just before his crucifixion, rode upon an ass 'as any prince or ruler would have done who was engaged on a peaceful journey.' 'Meek

and lowly was he, as became the new character hitherto unknown to the warlike and
restless Jews—a Prince, not of war, as had
been all other celebrated kings, but of peace.
Had he come as the Jews expected, despite
so many prophecies, their Messiah to come,
as a great king and conqueror, he might
have ridden the war-horse and been surrounded with countless legions of armed
men; but he came as the herald of peace,
and not of war, and, though meek and
lowly, yet a Prince, riding, as became a
prince, on an ass-colt which had borne no
inferior burden.'"

"I am glad, Miss Harson," said Malcolm,
"that you have read us that, because I always thought our Saviour rode upon an ass
to show how poor and humble he was."

"You will see," replied his governess,
"more and more, as we study the Bible
with the help of histories and travels, that
fully to understand it we must be well
acquainted with the customs and the ideas
of Oriental countries; and our talks upon
donkeys alone will help to improve our
knowledge of this subject. Here is a picture

of Syrian asses (p. 48) that will show
you what pretty animals they are.—These

THE ENTRY INTO JERUSALEM.

look like 'little horses,' Edie; they have
narrow heads, you see, instead of broad

ones, like other donkeys, and their ears are not nearly so large."

EASTERN ASSES.

"Aren't they white?" asked Edith.

"Perhaps they are, dear; for white asses

are found in the East, and very beautiful little animals they are. They are used only by persons of high rank, as they were in old times, for in the song of Deborah and Barak * we find the words, 'Speak, ye that ride on white asses, ye that sit in judgment, and walk by the way.' These white asses are larger and swifter than the ordinary ones, and with a handsome crimson saddle and a bridle of light chains and red leather such a steed makes a very fine appearance. 'Is that active little fellow,' writes a traveler, 'who with racehorse coat and full flanks moves under his rider with the light step and the action of a pony—is he the same animal as that starved and head-bowed object of the North, subject for all pity and cruelty and clothed with rags and insult?'"

"I should like to see the pretty saddles and chains," said Clara. "I wish they'd put some on our *horses.*"

"They'd frighten all the other horses then," suggested Malcolm, "and all the mad bulls would run after 'em."

* Judg. v. 10.

4

This was an alarming picture until Miss
Harson quietly said,

"How many mad bulls do you usually
see at once?"

The children laughed at the idea that had
suddenly entered their heads of a herd or
so of mad bulls running after the crimson
saddles, and they found that, like most ·
frights, it would not bear very close examination.

"The horses would be no more frightened than these Eastern steeds are," continued the young lady, " if, like them, they
were accustomed to gay trappings. But
such display, which seems natural to the
Orientals, would not suit our more sober
people.—The saddle and the bridle which
you admire so much, Clara, are quite different from ours, and the saddle especially is
a very complicated affair in two or three
pieces. First there is spread over the
animal's back a thick woolen cloth folded
several times; then comes a very thick
pad of straw covered with carpet and having a flat top. The pommel is so high that
the rider is perched ever so far above the

animal's back; and over the saddle a handsome cloth of bright colors hangs down at
the sides, and is often very costly. It is
edged with gold and colored fringe and
tassels, and the bridle also has tassels—
often bells, embroidery, and other ornaments. A first-class steed belonging to
a wealthy Syrian is really a gorgeous affair.
—Now, Malcolm," added his governess,
"you may read to us this description of
the ass in Egypt and Palestine."

Malcolm was a good reader for a boy of
his age, and he felt it to be quite an honor
to read several paragraphs to the little
company:

"'What, then, are the characteristics of
the ass? Much the same as those which
adorn it in other parts of the East—namely,
it is useful for riding and for carrying burdens; it is sensible of kindness, and shows
gratitude; it is very steady, and is larger,
stronger and more tractable than its Euro-
pean congener; its pace is easy and pleasant; and it will shrink from no labor if only
its poor daily feed of straw and barley is
fairly given. If well and liberally supplied,

it is capable of any enterprise and wears
an altered and dignified mien, apparently
forgetful of its extraction except when un-
deservedly beaten by its masters, who, however, are not so much to be blamed, because,
having learned to live among sticks, thongs
and rods, they follow the same system of
education with their miserable dependants.

"'The wealthy feed him well, deck him
with fine harness and silver trappings, and
cover him, when his work is done, with rich
Persian carpets. The poor do the best they
can for him, steal for his benefit, give him a
corner at their fireside, and in cold weather
sleep with him for more warmth. In Palestine all the rich men, whether monarchs
or chiefs of villages, possess a number of
asses, keeping them with their flocks, like
the patriarchs of old. No one can travel
in that country and observe how the ass is
employed for all purposes without being
struck with the exactness with which the
Arabs retain the Hebrew customs.

"'The result of this treatment is that the
Eastern ass is an enduring and tolerably
swift animal, vying with the camel itself in

its powers of long-continued travel, its
usual pace being a sort of easy canter.
On rough ground or up an ascent it is
said even to gain on the horse—probably
because its little sharp hoofs give it a firmer
footing where the larger hoof of the horse
is liable to slip.' "

Clara and Edith were particularly interested in the idea of sleeping with a donkey,
which they were quite sure they should
not like; but Miss Harson was equally
sure that if they had always been accustomed to it, like the poor children of the
East, they would not mind it at all.

"The wild ass," continued their governess, "is often mentioned in Scripture, and,
judging from the pictures of it, this also
seems to be a pretty little animal. In
summer he wears a gray coat with a red-
dish tinge, but in winter the red disappears.
He is wild in every sense of the word, and
swifter than the best horses, seeming to
see, hear and scent the most distant approach. No matter how young the wild
ass may be captured, it seems impossible
to tame him, and in the book of Job we

find this description: 'Who hath sent out
the wild ass free? or who hath loosed the
bands of the wild ass? Whose house I
have made the wilderness, and the barren
land' (or salt places) 'his dwellings. He
scorneth the multitude of the city, neither
regardeth he the crying of the driver. The
range of the mountains is his pasture, and
he searcheth after every green thing.'*
These animals assemble in herds, sometimes of several hundred, and wander in
search of 'every green thing' over large
tracts of country. 'Like many other wild
animals, they have a custom of ascending
hills or rising grounds and thence survey-
ing the country, and even in the plains
they will generally contrive to discover
some earth-mound or heap of sand from
which they may act as sentinels and give
the alarm in case of danger.'"

This seemed very comical to the children;
and if even wild asses could do this, it was
scarcely so wonderful that Balaam's ass
should speak.

* Job xxxix. 5–8.

CHAPTER III.

IN THE FAMILY.

"MISS HARSON," asked Malcolm, "is a mule the same as a donkey?"

"No," replied his governess; "a mule is half a horse and half a donkey, or ass. Mules are larger and stronger than donkeys, but have all their lightness, surefootedness and endurance. They are very desirable for carrying burdens over rough countries, and also for the saddle. The Andalusian mules, which are seen in Spain and the Spanish colonies, are very large and handsome, and in traveling among the Andes no other beasts can stand the risks and exposure which they bear with ease."

"I don't think there are any mules in the Bible," said Clara; "I don't remember seeing the name there."

"Then," said Miss Harson, smiling, "you

55

will have to find the thirteenth chapter of
Second Samuel and read us the last half of
the twenty-ninth verse."

Clara was very much surprised as she
read slowly the words,

"'Then all the king's sons arose, and
every man gat him up upon his mule and
fled.'"

"There is another place, a few chapters
on," continued the young lady, "where
a mule is mentioned in connection with a
dreadful event.—Can you not remember,
Malcolm, what it was?"

How Malcolm wished that he could! but
it seemed to the perplexed boy that the more
he tried to remember, the more he could not
think of it.

"I am sure, though, that you can remember, all of you, Absalom's wicked rebellion
against his father and his king, and how
dreadfully he was punished for it? It is
written: 'And Absalom met the servants
of David. And Absalom rode upon a
mule, and the mule went under the thick
boughs of a great oak, and his head caught
hold of the oak, and he was taken up be-

tween the heaven and the earth; and the mule that was under him went away.'"*

"But why didn't he fall," asked Edith, in a great puzzle, "when the mule went away?"

"Because, dear, his long, thick hair was, as we are told in another chapter, 'heavy on him,' and this became entangled in the tree, so that he could not escape. In those days mules were considered good enough for the king's sons, and even the king himself, to ride upon. In First Kings, first chapter, David commanded, 'Cause Solomon my son to ride upon mine own mule;' and this was considered the same as putting him on the throne."

"Are mules very obstinate?" asked Clara. "Jane told John yesterday that 'he was as obstinate as a mule.'"

Miss Harson could scarcely keep from smiling, for that worthy man's obstinacy was proverbial; but she replied quite soberly:

"There is nothing said about it in the Bible, Clara, which seems a little strange,

* 2 Sam. xviii. 9.

as travelers declare that the Eastern mules
are quite as perverse as those in other
countries. They say that 'they are very
apt to shy at anything or nothing at all;
they bite fiercely, and every now and then
they indulge in a violent kicking-fit, fling-
ing out their heels with wonderful force
and rapidity, and turning round and round
on their fore feet so quickly that it is hardly
possible to approach them.'"

"Well," said Malcolm, "I'm glad that I'm
not a king's son, if I should have to ride on
a mule."

"It does not sound attractive, certainly,"
said his governess, "but the royal mules
were probably too well trained to behave
in this way. Mules are famous all the
world over for kicking and for objecting
to do the particular thing that is required
of them. They often stand quite still when
it is particularly desirable that they should
go on; but when they can be cured of
these failings, they are very valuable animals. They carry heavy burdens with the
utmost patience, and find their way where
no horse could manage to pass. They can

HEAVILY LADEN.

appreciate kindness as well as other animals, and a traveler in Spain said that he
found his mule more tractable than many
reasoning bipeds.   This was because he
shared his food with him and talked to
him in a friendly, caressing way, 'when he
would wag his long ears as though con-
scious that some compliment was paid
him.'"

"Was not that cunning?" exclaimed
Edith. "It seems just like a dog."

"Mules are like dogs in desiring to be
well treated and caressed; and the more
they have of this kindness, the less obstinate they are. They always seem obstinate
and contrary with drivers who beat them to
make them go. Properly treated, the mule is
a good and faithful animal and a most valuable beast of burden. It is often employed
in our own country to draw heavy loads, and
in the days of canal-boats it was very useful. In Old-Testament days, also, the mule
was used for other purposes besides riding,
and Naaman said to the prophet Elisha,
after he had been cured of his leprosy,
'Shall there not then, I pray thee, be given

to thy servant two mules' burden of earth?'
These mules, however, were probably as
different from the others as our carriage-
horses are from the poor omnibus drudge.

"Another thing," continued Miss Harson, "for which these animals are valuable
is mountain-traveling; and among the Alps
and Andes there are fearful precipices which
only mules can pass in safety. Even they
would often rather not, but no other animal
will, make the attempt. There will be a
narrow path with immense heights on one
side and dreadful chasms on the other, and,
as though this were not bad enough, the
road itself will go down for several hundred yards at every little distance. The
mules are very cautious when they come
to these places, and they seem to be fully
aware of the danger they are in; so quite
of their own accord they stand perfectly
still on the edge of one of these descents,
and the driver may urge them on in vain
until they are quite ready to move. These
animals—so often called 'stupid'—are evidently considering just how that particular
precipice is to be managed. They look

carefully over the road and tremble at the
danger before them, but, all the same, they
go. They begin by placing their fore feet
as if stopping themselves; then they bring
the hind feet together rather forward, as
if they were going to lie down. The next
thing is a sudden slide down the declivity,
very much like 'coasting' down a hill on
the snow; but it is done in a flash, and the
rider has as much as he can do to keep
himself fast on the saddle, for the least
movement to one side or the other would
result in the destruction of both mule and
man."

"Oh!" said Clara, with a long breath; "I
should not like to travel in that way."

"I would," cried Malcolm; "it's perfectly
splendid. Just think of sliding down hill
like—like—"

"A mule," finished his governess, laughing.

Edith was too sorry for "the poor
mules" to see anything funny in it, and
she had never felt a desire to slide down
hill in any way.

"Now, Miss Harson," said Clara, coaxingly, "is not there some nice story about donkeys and mules?"

"There are some queer legends and fables," was the reply, "and—and I will see what I can do."

The children thought this quite equal to one story, at least, and they also wanted the legends and fables, and all that was to be had.

"As to the legends," continued the young lady, "one of them, which is very ancient, refers to the black mark which the ass has along the spine and across the shoulders, forming a complete cross. This mark is said to have been placed upon the animal as a memorial of our Saviour's riding upon an ass on his triumphal entry into Jerusalem. 'There is another Christian legend respecting the ass of Palestine, which is thought to owe its superiority in size, swiftness and strength to the fact that it helped to warm the infant Saviour in the manger, that it carried him and his mother into Egypt and back again, and that it was used by the Lord himself and his disciples.' A legend of the mule says that 'when the

holy family were about to travel into Egypt,
Joseph chose a mule to carry them. He
was in the act of saddling the animal when
it kicked him, after the fashion of mules.
Angry with it for such misconduct, Joseph
substituted an ass for the mule, thus giving
the former the honor of conveying the family into Egypt, and laid a curse upon it that
it should never have parents nor descend-
ants of its own kind, and that it should be
so disliked as never to be admitted into its
master's house, as is the case with the horse
and other domesticated animals.' "

Little Edith was listening with wide-open
eyes, while Clara said,

"It isn't any of it true, is it, Miss Harson?"

"No, dear," was the reply, "we have no
reason to suppose that it is; yet many such
ideas originated with pious but ignorant
people. I think you will like some fables
better, and to understand the first one you
must know that the horse has always been
supposed to have a great contempt for the
ass, and to dislike him because of his strong
resemblance to himself.

"A man who owned both a horse and an ass gave the latter animal nearly all the work to do and loaded him with burdens, while the horse would canter off with a grand, independent air, having nothing to carry unless it was the master himself, and this he thought an honor. As for that despicable ass, he reflected, he was glad that he was not in his place, but that was because he was a handsome animal and a credit to his owner.

"One day the poor ass was so terribly overburdened that he felt his strength giving out, and, as the horse was not even carrying his master, he appealed to him for help.

"'If you will kindly take part of my burden,' said he, 'you will not mind it, and I shall be able to get on; but unless you divide it with me, I shall sink down exhausted.'

"The horse's only reply was a snort of contempt, and his humble companion pleaded again and again without any success. Then he fell down under his burden, and died; and the horse, who would not

make the least effort to save him, was
properly punished, for no sooner did their
owner see what had happened than he took
the load from the back of the ass and piled
it all on the horse; then, wishing to use the
ass's skin, he slung the dead animal also
over the disgusted living one, who, for
having refused to carry part of his companion's load, was now obliged to carry the
whole of it and the ass himself besides."

The children were delighted with this
speedy and just punishment, and their
governess told them that it was one of
Æsop's fables which she had repeated to
them—not as it was in the book, but as she
remembered it.

"There is another one," she continued,
"of an ass who overreached himself very
much as the horse did. He was owned by
a huckster who took him quite a long distance to the seashore to get salt for his
customers, as he had been told that it was
much cheaper there. A very liberal load
was piled on the donkey, and the animal
did not enjoy his burden at all. In passing
a slippery ledge of rock on their way back

to the town the ass fell into the water below, and the salt, of course, was melted. As the ass was not hurt, however, he thought it a fortunate tumble because it had rid him of his load, and he went home rejoicing.

"But soon after the huckster tried it again; and when they reached the sea-shore, the ass was even more heavily loaded than before. This was not to be borne; and when they reached the stream where he had gotten rid of his former burden, down went the donkey, salt and all, into the water. But his master thought this to be quite overdoing the matter, and the next time the slippery animal was provided with a load of sponges instead of salt. When he tried his old trick with these, he found that, instead of getting rid of them, the water had made them twice as heavy as they were before. The journey home seemed a very long one, and he had plenty of time to reflect that his own cunning had brought him into this unpleasant predicament."

"I wonder if that is true?" said Malcolm, forgetting that it was a fable.

"Something very much like it might be," was the reply, " for I think that animals have much more sense and cunning than is generally supposed. But have you not all had enough fables?"

Not if there were any more, and Miss Harson laughingly insisted that she should stop with an account of the ass that wanted to be a lap-dog. This promised to be very funny, but the young lady warned her audience that it had a sad end:

"There was once a donkey who was a well-behaved animal, and certainly very well off as donkeys go. He had comfortable quarters, plenty of good food and was never overworked. He was fond, too, of his master, and liked to have him come to the stable now and then and pat him. This was all he ever expected, and he was quite happy and contented until a wretched little lap-dog came into the family—that is, he called it 'wretched' because he did not like it, but it was really a very pretty, sprightly little animal. What a fuss his master did make over that dog! He was always teaching it tricks and laughing at its an-

tics, and he would actually let it curl itself
up in his lap and go to sleep there. He
never treated his faithful donkey with so
much favor, and the affectionate but ugly
animal set himself at work to find out the
reason. First he would watch that miserable little dog to see exactly what he did,
and then go and do it himself; for he had
probably been too quiet and had stayed too
much in the background. Great was the
consternation one day when the ass, having
broken from his fastenings in the stable,
came prancing into the house and presented himself to his master rearing up on
his hind legs, as he had seen the little dog
do. Then he rolled over and over on the
floor in imitation of 'Flip,' and playfully bit
and tore at things, as that small animal
sometimes did. He had already knocked
down a table and broken the valuable china
on it, chewed some handsome books and
torn the curtains before the servants could
get at him, for fright; but when he put his
hoofs on his master's lap and prepared to
vault into it like the lap-dog, he was attacked with brooms and cudgels on all

sides as a mad creature, and beaten so
unmercifully that he never got out of the
room alive. He was a good donkey, but
a very undesirable lap-dog."

This story was pronounced the best of
all, and the children laughed merrily at the
idea of a great donkey acting like a little
dog. It was too bad, though, that people
did not understand what he meant by it,
instead of killing him; and if he had not
been killed, he would have been ever so
sorry that he did not stay in his stable.

This last was from Edith, whom Malcolm
pronounced "a dreadfully funny little thing,"
and whose remarks certainly were rather
mixed up. But the little girl herself could
not see that she was at all " funny."

" Now," said Miss Harson, "are there
any children so insatiable as to want a
regular story after all this? Because, if
there are "—the pause was rather alarming—" why, I've got one ready for them."

The speaker was so desperately hugged
for the next few minutes that she laughingly declared it was almost as bad as
encountering a party of bears. No; it

was not one of her own stories, which
would not have been nearly so good—here
there was a series of incredulous looks—
but was a very bright one written by Mr.
W. H. Beard for *Harper's Young People;*
and, having finally recovered her breath,
the young lady began to read

THE WAYWARD DONKEY.

There was once a little donkey who
gave his poor mother no end of trouble,
he was so stubborn, unreasonable, exacting
and dreadfully saucy. Why, when angry
he didn't hesitate at all to call his mother
an old donkey right out. One day, when
crossed in some particularly absurd desire,
he declared he would run away. Immediately putting his threat into execution,
off he trotted, heedless of his poor fond
mother's entreaties. Away he went, sus-
tained at first by his temper and pride, but
as the day wore on he became weary, faint
and hungry. The matter of food and shelter became a question of serious alarm, and
how to obtain them was a problem too great
for his little donkey-brain to solve. He

now remembered that he had never had
to trouble himself with all this before, all
the needs and comforts of life having been
provided for him without thought or care
on his part.

The land over which he was traveling
was quite poor and afforded only a few
little stunted thistles, which seemed to consist more of prickers than anything else,
which pierced his tender little nose and
made it bleed.   He saw plenty of oats and
other grains, as well as nice vegetables,
growing in fields, but so well guarded by
high fences that he could not hope to get
at them.   Many times, when hunger and
fatigue had subdued his pride, would he
have returned home; but he had wandered so far that he had not the least
idea which way he had come.   To add to
his distress, he saw the sun was fast declin-
ing; already he felt the chills of evening.
But there was no use bemoaning his fate,
and he must make the best of it.

At length, too weary to travel farther, he
was forced to lie down to rest, and selected
for the purpose an unfenced, overgrown

piece of ground of considerable extent. Here, as he lay among the weeds, nothing was visible of him above their tops but his two ears, which might easily have been taken for two stakes or the roots of an upturned stump. As he lay shivering in the damp grass he felt anything but comfortable. The sun went down; the moon arose and shed a cold light over the face of nature which made him feel lonely indeed.

Suddenly there appeared above the grass several other pairs of ears bobbing about quite like his own. The sight thrilled him with something akin to pleasure, for he asked himself, "To whom can such ears belong but to little donkeys? and if young donkeys are around, they must have mothers, or a mother, near by, who, no doubt, would be very glad to adopt such a fine specimen of the race as I." (The reader has already seen that he was a conceited little donkey.) So saying, he arose quickly to his feet. The others stood up also— though not, as he did, on their four feet, but on their hind legs; that is to say, they stood up on their haunches—and looked at him in blank amazement; but as he approached them they bounded away so fast that it was useless to try to overtake them. When he stood still, they also stopped and again stood up on their haunches and peered at him over the tops of the weeds. Master Donkey did not try again to go to them, but expostulated with them upon their ill-breeding and unkind behavior, called them cousins, told them he was tired and hungry and asked for food and shelter.

This touched their tender little hearts, and they cautiously drew near and made the acquaintance of their supposed cousin. On a close scrutiny, however, they doubted his claim to relationship, and flatly told him so. But they good-naturedly said if he was hungry it was no more than common humanity first to relieve his wants, and discuss the question afterward: even murderous man would do as much as that; so they brought him carrots and other vegetables in abundance from a farm-garden near by from which they were accustomed to supply their own wants.

When his appetite was satisfied, his humanity, such as it was, oozed out, and he became as arrogant as ever, and stoutly claimed that he was their big cousin—though, he said, he was not particularly anxious to be acknowledged by such a pack of little dwarfish, thieving creatures as they were, who would steal through the farmer's fence to purloin vegetables for a cousin whom they impudently refused to recognize. Their spokesman retorted and said they claimed a right to a share sufficient for their needs of whatever grew upon the earth. To be sure, they were obliged to obtain it stealthily at night, as the man claimed it all for himself and it would be almost certain death to be found by him within his enclosure; indeed, many of their unfortunate fellows had already suffered death for the exercise of this natural right. If, however, he regarded their act as a crime, he was himself a criminal, inasmuch as he had accepted the fruits and profited by the act, knowing how the fruit had been obtained. To this the donkey could make no answer; at least, he did not think it prudent to try, as night was still before him and the question of shelter still unsolved.

Good-nature was soon restored and the discussion renewed. The rabbits ("I thought so!" exclaimed Malcolm) could see many points of difference, but only two of resemblance. It certainly could not be denied that the ears were remarkably like and the complexion was very nearly the same, but the hard feet were so widely different from their own soft paws. And the tail, too—long and dangling, like a cow's! What a tail for a rabbit! Then, again, they had observed that he stood while eating, whereas a true rabbit always crouched comfortably near the ground while taking his food. In the matter of voice, too, they flattered themselves there was a wide difference. However, all this might be changed or improved by judicious training, except the feet; the hoofs they despaired of. The tail they proposed to nibble off at a proper length from the body. This operation the donkey positively refused to submit to, but finally consented to hold his tail up over his back, as much like a rabbit as possible, and, moreover, would at once set about his lessons to learn their ways, so that he might the sooner adapt himself to their habits and become one of them.

Accordingly, one of the cleverest of their number was charged with his instruction, and immediately began with the important art of sitting on the haunches with his tail curled up upon his back. In this, though he strained every nerve to perform it, he made an ignominious failure. He could maintain the position for only a moment, and then would pitch forward or fall backward, seeming to rock over on his curved tail and cutting such a ridiculous figure that it made all the rabbits laugh. This made him very angry, and he began to use his heels in a most vigorous and unrabbitlike manner. All ran for their lives, but not all escaped unhurt. The "spraggly" forms of two or three of them nearest him showed dark against the moonlit sky before they limped off and, joining their fellows, gathered in a little knot at a distance from their fractious pupil and discussed his merits with

great freedom. They voted him an ill-
natured brute, a stupid dolt—in short, a
perfect donkey. Scarcely had they arrived
at this unanimous conclusion, when "Pop!
pop! Bang! bang!" four loud reports,
and four little rabbits lay in the agonies
of death.

The farmer and his son, seeing by the
moonlight strange movements in the field,
had stolen upon them with their double-
barreled guns in the unguarded moment
of their excitement, and, as the boy expressed it, bagged four rabbits and a
donkey; for poor little donkey stood paralyzed with fear. He had never looked upon
death before, and was an easy captive.
Without troubling himself to inquire who
was the rightful owner, the farmer took
him for his own and housed him that night
in a stall by himself, where he passed almost
the entire night, notwithstanding the fatigues
of the day, in such reflections as he was
capable of.

He grew up to be a great donkey, to be
sure, but the lessons of that day were never
forgotten by him.

The children were so much interested in this story that it was quite a shock to have it end so suddenly, and they asked in a very injured way why they were not told how the donkey got back to his mother. Their governess could only reply that she did not know; perhaps he never got back at all.

CHAPTER IV.

WITH A HUMP.

"I SHOULD like," said Miss Harson, "to have some one, or all, of you tell me what other animal is quite as much a beast of burden as the ass, and quite as useful to man—even more so in hot or dry countries."

"Is it in the Bible?" asked Malcolm.

"Yes," was the reply; "it is a Bible animal, and you will see its portrait in most books of Eastern travel."

"I know!" cried at least two voices. "It's the camel!"

"The very creature, and in many respects it is one of the most interesting and wonderful of animals. It is called the 'ship of the desert' because it safely navigates the sandy sea where other animals would perish beneath the burning sun, and carries the traveler, too, 'to the haven where he would

LOADED CAMELS.

be.' The Arabs value their camels very highly, and it is well for them that they can find sufficient food in the thorny bushes which are found here and there in the desert, as there is not much else to be had."

"What funny-looking things they are!" said Edith, who was gazing with great interest at a picture of some loaded camels which Miss Harson had just shown her in a book. "They are so dreadfully high! Don't the people have a ladder to climb up on their backs?"

"No, dear; I never heard of a camel ladder, but I quite agree with you that they do not look like comfortable animals to ride on."

"There's the hump, too," said Clara— "or two of 'em. Some seem to have one, and some two. Why is that, Miss Harson?"

"Because they are different kinds of camels. The Arabian camel has but one hump, while the Bactrian camel has two. The dromedary is also a camel with one hump, as it is only a finer breed of the

Arabian species. It goes faster than the ordinary kind, and its name means 'a runner;' its principal use is to carry messages across the desert. Here is a picture of one with a postman on his back, and you can see how fast the animal's legs are moving."

"What awful legs they are!" said Malcolm. "No wonder the camel looks as if he was up on stilts when he is standing straight."

"Will he bite?" asked Edith, with a shrinking from even the mouth in the picture. She did not think it had an amiable expression.

"I am sorry to say," replied her governess, "that it has been known to do so when angry; and as its teeth are very sharp and strong the better to cut up its hard, thorny food, and as it can twist its long neck about so quickly, it is not advisable to get within biting-reach."

"I don't like its neck and mouth," said Clara; "they put me in mind of a big snake."

"I don't like any part of it," added

Malcolm. "I think that even a donkey is prettier."

"What does it have such a short tail for," asked Edith, "when it's such a great big creature?"

"That is the latest style, dear—for camels," said Miss Harson, who was quite amused at the unpopularity of the grotesque-looking animal. "They never have sweeping tails like horses, nor even like donkeys. These funny creatures are very useful in their native countries, but, unlike the horse and the ass, they are of no use anywhere else."

"But how," asked Malcolm, in a great puzzle, "does any one ever climb up on a camel to ride him?"

"The rider does not climb up at all, as I meant to have told Edith," was the reply, "but the camel kneels to receive its burden. It first drops on its knees, then on the joints of the hind legs, after that on its breast, and then again on the bent hind legs. 'Kneeling is a natural position with the camel, which is furnished with large callosities, or warts, on the legs and breast, which act as cushions on which it may rest its great weight without breaking the skin.' So you see that, like many other animals, it is wonderfully fitted for the uses to which it is put."

"How tall is a camel, Miss Harson?" asked Clara. "It looks almost as high as a house."

"Not quite, dear," was the smiling reply, "but, as a tall camel will measure seven feet from the ground to the top of the hump, and the saddle with its cushions adds a foot or two more, it is easy to understand that a fall from such an animal's back is not a trifling matter."

"I should think, though," said Malcolm, who was intently studying the ungainly creature, "that it would be about as pleasant to fall off as to stay on. How does any one manage to stick on a point like that?"

"You must remember that no one tries to ride on a camel without a saddle, and the saddle makes a great difference. Besides the cushions with which it is provided, there is in front a long upright piece to which the

rider can cling to prevent his being thrown
off. The Arabs pass one leg over this
wooden peg and hitch the other leg over

THE CAMEL.

the foot that hangs. But the safest way
of sitting, though not the most comfortable,
is to cross the legs in front and grasp the
pommel with both hands."

"I shouldn't think," said Clara, "that any of it would be 'comfortable.'"

"It wouldn't suit our ideas of comfort," replied Miss Harson, "and I am sure that three little people of my acquaintance, with their governess, would be dreadfully seasick if they tried this style of riding."

Seasick on a camel! The little girls could not understand this.

"Of course," said Malcolm, with a knowing air; "it's a ship, you know—the ship of the desert."

"It is certainly a ship in the way of seasickness, according to the experience of travelers when they are learning to ride. The reason of this is that the movement of the camel is entirely different from that of any other animal used for this purpose, as it moves a front leg and a hind leg together on one side, and then those on the other. This makes a long, swinging motion which is like the pitching of a ship on the sea. When rising from the ground to begin this trot—if it can so be called—the animal suddenly straightens its hind legs first, which jerks a rider who is not pre-

pared for this movement directly over its
head. An English traveler says that any
one who wishes to practice camel-riding at
home can imitate it admirably 'by taking a
music-stool, screwing it up as high as possible, putting it into a cart without springs,
sitting on the top of it cross-legged, and
having the cart driven at full speed trans-
versely over a newly-ploughed field.'"

The children laughed at this, but they
did not seem inclined to try the experiment,
and Miss Harson laughingly declared that
she had no curiosity on the subject.

"Travelers agree," continued the young
lady, "in saying that there is as great a
difference in the gait of camels as in that
of horses, but the movement even of the
'smooth-going ones' is very hard at first,
'causing the body of the rider to swing
backward and forward as if he were rowing
in a boat.' It makes his back ache dread- ·
fully, besides the feeling of seasickness,
and it is so stiff the next day that he cannot
move without suffering severely. An Eastern traveler says that he tried to sit up
straight on his camel without moving, and

for a few moments this seemed to be a
relief; but, getting very tired of this
position, he next tried lying down with
his head resting on his hand: 'This last
manœuvre I found would not do, for the
motion of the camel's hind legs was so
utterly at variance with the motion of his
fore legs that I was jerked upward and
forward and sideways, and finally ended
in nearly rolling off altogether. Without
going into the details,' he adds, 'of all that
I suffered for the next two or three days—
how that on several occasions I slid from
the camel's back to the ground, in despair
of ever accustoming my half-dislocated
joints to the ceaseless jerking and swaying
to and fro, and how that I often determined
to trudge on foot over the hot, desolate
sand all the way to Jerusalem rather than
endure it longer—I shall merely say that
the day did at last arrive when I descended
from my camel, after many hours' riding, in
as happy and comfortable a state of mind
as if I had been lolling in the easiest of
arm-chairs.' "

Malcolm began to think that he would

like to try camel-riding, after all, and his
governess said that he should—as soon as
he became an Eastern traveler. This did
not seem very near, and it comforted his
little sisters for the anxiety they were beginning to feel on his account.

But the boy was still interested in the
camel as a possible steed, and, looking at a
picture, he said,

"How queerly his bridle is put on!—
right under his eyes, and not in his mouth
at all."

"It is not a bridle," replied Miss Harson,
"but a nose-string, which is a rope tied like
a halter round the muzzle, with a knot on
the left side. The rider holds it in the left
hand and uses it for the purpose of stopping the animal. 'The camel is guided
partly by the voice of its rider and partly
by a driving-stick, with which the neck is
lightly touched on the opposite side to that
which its rider wishes it to take. A pressure of the heel on the shoulder-bone tells
it to quicken its pace, and a little tap on the
head, followed by a touch on the short ears,
is the signal for full speed.'"

TAKING A REST.

"This nice camel," said Clara, "in the picture of 'The Camel-Post,' has got a queer blanket on and things hanging down from it."

"That is his saddle, dear, which on a fine camel is of rich material ornamented with handsome fringe and embroidery. The ugly hump does not show so much when it is covered with such a saddle."

"What is his hump made of?" asked Edith. "Does it grow right out of his bones?"

"No, Edie; it has nothing to do with the bones, as it seems to consist of fat, and the animal is supposed to absorb this fat when it is obliged to go a long time without food, as the hump is then very soft and flabby. When a camel is in good health and well cared for, this hump is high and firm; so that it is easy for those who understand these animals to judge of their condition by that of the singular-looking hill on their backs."

"What a funny foot the camel's got!" said Malcolm. "It looks like two great toes."

"That is just what it is," replied her governess, "for it is described as 'two long toes resting upon a hard, elastic cushion with a tough and horny sole. This cushion is so soft that the tread of the huge animal is as noiseless as that of a cat, and, owing to the division of the toes, it spreads as the weight comes upon it, and thus gives a firm footing on loose ground. The mixed stones and sand of the desert would ruin the feet of almost any animal, and it is necessary that the camel should be furnished with a foot that cannot be split by heat like the hoof of a horse, that is broad enough to prevent the creature from sinking into the sand, and is tough enough to withstand the action of the rough and burning soil.'"

"How wonderful it seems," said Clara, "that he should have just what he needs!"

"'His providence is over all His works' are words of which we are constantly reminded in our studies of Nature. Nothing has been overlooked nor deemed too trifling by the great Creator in caring for his creatures, and a camel never seems to hurt its feet in walking over sharp stones or thorns

or roots of trees.   But whenever it happens
on mud, where it does not belong, it seems
incapable of walking at all.   Here 'it slips
and slides, and generally, after staggering
about like a drunken man, falls heavily on
its side.'   Neither does it like deep, loose
sand, but groans at every step, as it is very
tiresome to drag its feet out of the holes
into which they sink."

The children thought this "very queer,"
but Miss Harson told them that there was
something else about the camel that seemed
still queerer.

"This ungainly animal," she continued,
"whose life is chiefly spent in the desert,
where food and water are not often to be
had, has the singular power of going without both for an almost incredible time; and
when water is plentiful, it will at once drink
a great quantity that lasts for several days.
The stomach has a number of cells, into
which the water runs as fast as it is swallowed, and as much as twenty gallons will
be taken at once, and disposed of afterward
by slow degrees."

This was certainly the strangest thing the

CAMELS IN THE DESERT.

children had ever yet heard about the camel,
and now they sat waiting for fresh wonders.

"I do not know of anything quite equal
to that," said the young lady, "but I think
that the camel's food is little short of a
marvel. The idea of its getting nourish-
ment from thorn-bushes—thorns and all—
is very puzzling, particularly as the thorns
are an inch or two long and as sharp as
thorns can be. But the animal's palate is
hard and horny, and seems to take kindly
to this unpromising diet. It does not matter how dry and withered the twigs are: the
camel enjoys them as much as some other
animals do fresh vegetables; and some one
has said that this strange creature could
thrive on the shavings of a carpenter's shop."

"I wonder if he would eat 'em?" said
Edith, quite seriously, and Malcolm promised her that as soon as he found a camel
near a carpenter's shop he would try him
with some shavings.

"There is something else about these
thorn-bushes," continued Miss Harson,
"which I should like to have you remember. Owing to the great heat and dry-

ness of the climate where they grow, they
blaze up in a moment, if lighted, with a roar
and crackling, and disappear in a flash.
The prophet Ezekiel says, 'For as the
crackling of thorns under a pot, so is the
laughter of a fool,' these thorns being
used as fuel in the desert. Far back in
the Old Testament there is an account of
a miracle in connection with one of these
thorn-bushes.—Can you remember it, Malcolm?"

"Was it Moses and the burning bush,
Miss Harson?"

"Yes. When Moses was at the back of
the desert with his father-in-law's flock, 'the
angel of the Lord appeared unto him in a
flame of fire out of the midst of a bush:
and he looked, and behold, the bush burned
with fire, and the bush was not consumed.
And Moses said, I will now turn aside, and
see this great sight, why the bush is not
burnt.' It was so wonderful a thing that
the light, dry thorn-bush should be in a
blaze, and yet continue blazing instead of
being reduced to ashes in a moment, that
Moses, evidently not thinking of a miracle,

7

as he had not yet seen the angel, turned aside to find out the reason. And this gives us some idea of the kind of food that camels live on in the desert."

"Don't the poor things have anything else to eat?" asked Clara, pityingly.

"Oh yes," replied her governess; "they have oats and barley when at home, and while traveling out of the desert they find a shrub called the *ghada* in which they particularly delight. This bush is often six feet high and not unlike a small palm tree, as it has a feathery tuft of little green twigs which are very slender and flexible. No matter what the hurry may be, it is almost impossible to get a camel past one of these bushes until he has gnawed off the top, and any amount of punishment will not prevent him from stopping again at the very next *ghada* he sees."

"Miss Harson," asked Edith, "what color are camels? Are there any white ones?"

"Yes, dear," was the reply; "there are white ones, and the color seems to vary in different places. Very often it is a sort of terra-cotta color, which, you know, is be-

tween red and yellow. Gray is also common, and there are a few, but not many,
black ones."

"Where are their ears?" asked Malcolm.
"They don't seem to have any in this picture."

"They do not make much show at any
time; and when the animal's full face is
turned, they are scarcely visible at all.
But, such as they are, you will find them
in the place where ears ought to be. There
is a legend that the camel once had very
long ears, and, being much dissatisfied with
its appearance, it asked for long horns too,
to balance them. But, instead of getting
horns, its ears were taken off almost close
to its head."

This sounded very funny, but it could not
have been pleasant for the camel.

"I wonder if little camels are pretty?"
said Clara.

"The young animal is described as
almost so in comparison," replied Miss
Harson, "but it is a funny, helpless little
object, and at first has to be watched like a
human baby. 'It cannot stand alone; with-

out help it cannot so much as take its own food, while its long neck is so flexible and fragile that, unless some one were constantly at hand to watch, the poor little creature would run every risk of dislocating it.' A little camel, it seems, does not play and gambol like other young creatures, but is just as grave and quiet as the grown-up ones, and it looks as melancholy as though it could see all the loads it would have to carry during its life. At three years old it begins to work, and it is trained to kneel and bear burdens, which are made heavier by degrees until it is eight years old. A camel is then quite grown up and can carry all it will ever be able to bear."

CHAPTER V.

WHAT THE BIBLE SAYS.

MISS HARSON had told the children on Sunday afternoon to look through their Bibles for places where camels were mentioned, and they produced quite a store of them when Monday evening came.

Clara had found the first mention in Gen. xii. 16, where it is said of Abram that "he had sheep, and oxen, and he-asses, and men-servants, and she-asses, and camels."

"Did Abraham ride on camels?" asked Malcolm.

"No doubt he did," replied his governess, "and one of our favorite writers says that 'Abram needed camels not only for their milk, and, for all we know, for their flesh, but for their valuable use as beasts of burden, without which he never could have traveled over that wild and pathless land.'"

"Why, Miss Harson!" exclaimed Edith, in great surprise; "do camels give milk? Horses don't."

"Yes, dear; camels do give milk which is very valuable to the Bedouins, or people of the desert. It is considered better when it is in a curdled state than when it is fresh, and from it is made a kind of salt cheese of which travelers do not speak very highly. Camel's-milk butter is churned in a different way from ours, as the milk is poured into a skin bag, and the bag is then beaten with a stick until the butter appears."

"I shouldn't want any of that butter," said dainty Clara.

"Not while you can get the golden prints that Kitty turns out," was the laughing reply; "but wait till you come to travel in the desert, and even this not very temptingly made butter will seem a luxury, as it does to the Bedouins.—But what have you found, Malcolm?"

"I've found about Rebekah at the well," was the somewhat disappointed reply, "but they're just the same old camels that Clara had."

"Abram's camels, I suppose you mean?" said his governess, rather gravely. "But that does not make them any the worse, as they are mentioned in another place. Tell us, please, where you found them."

"In Genesis, chapter twenty-four, verses ten, nineteen and twenty," read Malcolm from the list he had written.

"'It was by the offering of water to these camels,'" said Miss Harson, "'that Rebekah was selected as Isaac's wife;' and when, a great many years later, their son Jacob left his father-in-law, camels were among his valuable possessions, as we are told in the forty-third verse of the thirtieth chapter of Genesis."

"I know something about Job," said Edith, exultingly: "he had lots of camels—six thousand."

"And where does it say that, dear?"

"In the last chapter—Clara found it for me—and it's the—the twelfth verse. I can't remember all the other things he had."

"'Fourteen thousand sheep,'" read her governess, "'and six thousand camels, and a thousand yoke of oxen '—that means two thousand—'and a thousand she-asses.' "

The children looked aghast at the idea of all these animals, and they felt very glad that the beasts were not "around" at Elmridge.

"There was plenty of room for them," continued the young lady, "where land was so abundant and people were scarce, but here and now they would be rather in the way."

"Here is another verse, Miss Harson," said Clara : "'And he saw a chariot with a couple of horsemen, a chariot of asses, and a chariot of camels.' "*

"And here is a picture of 'A Chariot of Camels.' See what queer-looking camels they are."

"Their backs are right up in two hills," said Malcolm, "and they seem to have hardly any heads at all."

"These are the Bactrian, or Persian, camels," continued the young lady, "and they are generally used for drawing vehicles. They are much slower animals than

* Isa. xxi. 7.

the Arabian camels, and seldom get on faster than two miles and a half in an hour. But they are very hardy and do not mind the cold, walking on ice with the greatest ease. They are said to climb rocks so well that they are even more surefooted than mules; and the foot of this camel has extending beyond the other part a toe which forms a sort of claw and prevents it from slipping. It never requires any kind of shelter, and as it is left, even in the coldest kind of weather, to find its own food, it is not an expensive animal to keep. What Bible verse comes next?"

"Here is a long one," replied Malcolm: "'The burden of the beasts of the south: into the land of trouble and anguish, from whence come the young and old lion, the viper and fiery flying serpent: they will carry their riches upon the shoulders of young asses, and their treasures upon the bunches of camels, to a people that shall not profit them.'"*

"'Bunches,'" said Miss Harson, "means, of course, 'humps;' and it is surprising to

* Isa. xxx. 6.

find how much and what various kinds of
things are carried upon these bunches."

"But how does anything that isn't alive
ever stay on?" asked Clara. "I should
think the 'treasures' would all roll off."

"The strangely-shaped back with a small
mountain in the middle is provided with a
pack-saddle which keeps the burden in its
place, and this is the way it is managed: 'A
narrow bag about eight feet long is made
and rather loosely stuffed with straw or
similar material; it is then doubled and
the ends are firmly sewn together, so as
to form a great ring. This is placed over
the hump, and makes a tolerably flat surface. A wooden framework is tied on the
pack-saddle, and is kept in its place by a
girth and a crupper. The packages which
the camel is to carry are fastened together
by cords and slung over the saddle. They
are connected only by those semi-knots
called 'hitches,' so that when the camel is
to be unloaded all that is needed is to pull
the lower end of the rope, and the packages fall on either side of the animal. So
quickly is the operation of loading per-

formed that two or three experienced men can load a camel in very little more than a minute.' "

" Can a camel carry a great deal at once, Miss Harson?" asked Malcolm.

" An ordinarily large and strong one will carry from five to six hundred pounds on a

CAMELS AND TENT.

short journey, and about half as much on a long one. There is an Eastern proverb which says, 'As the camel, so the load;' and it is a remarkable thing about this animal that it knows just how much it can carry, and if the load is more than this amount it refuses to stir until it is relieved of the extra weight. But when properly started, it keeps on for hours at the same pace, and seems as fresh at the end of the journey as when it started. It objects, however, to being loaded at all, and does not resign itself to custom, like other beasts of burden, but growls and groans, and even tries to bite. 'So habitual is this conduct that if a kneeling camel be merely approached and a stone as large as a walnut be laid on his back he begins to remonstrate in his usual manner, groaning as if he were crushed to the earth with his load.'"

"What a humbug!" exclaimed the children.

"He even cries—for camels *can* cry—and it is quite affecting to see his piteous expression and the tears streaming from his eyes. But the drivers do not mind these in the least, for they understand the animal and know that they are only 'crocodile' tears. Sometimes the camel carries huge panniers in which an immense number of things can be stowed away, and a traveler speaks of seeing an Arab family traveling in this style. 'The wife and child,' he says, 'came by in this string of camels, the former reclining in an immense circular box stuffed and padded, covered with red cotton and dressed with yellow worsted ornaments. This family nest was mounted on a large camel. It seemed a most commodious and well-arranged traveling-carriage, and very superior as a mode of camel-riding to that which our *sitteen** rejoiced in, riding upon a saddle. The Arab wife could change her position at pleasure, and the child had room to walk about and could not fall out, the sides of the box just reaching to its shoulders. Various jugs and skins and articles of domestic use hung suspended about it, and trappings of fringe and finery ornamented it.'

"In the book of Judges," added the young lady, " it is written that when Gideon had slain the kings of Midian he 'took away the ornaments that were on their camels' necks.' † These ornaments, with which wealthy riders were fond of decorating their camels, were 'like the moon,' or crescent-shaped, made of silver and gold,

* Lady.   † Judg. viii. 21.

and would jingle at every step that was
taken. 'The chains that were about their
camels' necks' were so valuable that they
were classed with the 'ornaments, and col-
lars, and purple raiment that was on the
king of Midian.'"

"Miss Harson," asked Clara, "is there
anything in the Bible about the kind of
camels that carry letters?"

"Yes; 'the dromedaries of Midian and
Ephah'* are mentioned, and in Jeremiah
it is written, 'See thy way in the valley,
know what thou hast done: thou art a
swift dromedary.'† Also in Esther we
read that Mordecai 'sent letters and posts
on horseback, and riders on mules, camels,
and young dromedaries.' ‡ You will remember my telling you that dromedaries
are the handsomest and choicest kind of
camels, and the *deloul,* or post-camel, is
the best and fastest of these dromedaries.
With a light load the *deloul* will travel nine
or ten miles an hour, while the ordinary
camel usually makes but three miles in
that time. It will keep this up, too, for

---

* Isa. lx. 6.   † Jer. ii. 23.   ‡ Esth. viii. 10.

seven or eight weeks, with only a few
days of rest. But it is very hard on the
rider, and an Arab who can keep on a
deloul for a whole day boasts of it as a great
exploit. The camel-express messenger is
well known in India; and 'if it were not for
his long, quick rides on a rough, jolting
camel we might think he was the most
comfortably situated of all the postmen
He wears a red uniform and a large green
turban embroidered with gold thread. From
his belt hangs a curved sword in a red
sheath. The camel has trappings of gay
cloth and tassels ornamented with blue
beads and cowrie-shells, and small bells
are hung around his neck. Two heavy
mail-bags hang one on each side of the
camel, and the saddle is arranged to carry
a passenger behind the postman; but few
passengers care to travel that way, for the
jolting of a hurrying camel is so painful
that the poor postmen do not live very
long.'"

The little Kyles declared this to be a
shame, and thought that people had better
even do without their letters.

"It does not seem to be so bad in Arabia," continued their governess, "for there the postman prepares himself for his hard day's work by belting himself tightly with two leather bands, one just under his arms and the other round the pit of his stomach; this prevents him from suffering any serious injury. But at its best one would not expect the office of camel-postman to be much in demand."

"Miss Harson," said Malcolm, earnestly, "here is a verse from the New Testament about a camel that I do not understand a bit: 'Again I say unto unto you, It is easier for a camel to go through the eye of a needle, than for a rich man to enter into the kingdom of God.'* Why, a camel couldn't even begin to go through the eye of a needle!"

"I have seen a picture, though, where he is trying, and there is every reason to suppose that he will get through in time, and the others after him."

The children were very much surprised to see in the picture in the book which

* Matt. xix. 24.

Miss Harson held a kneeling camel with his hump in the air and his head through a very small doorway, while the driver urged him on with a stick, and other camels, with their drivers, were awaiting their turn. Malcolm read at the bottom of the picture, "A Camel going through a Needle's Eye," and under that were the words of our Lord which had so puzzled him.

"This is the description: 'In Oriental cities there are in the large gates small and very low apertures called, metaphorically, "needles' eyes," just as we talk of certain windows as "bulls' eyes." These entrances are too narrow for a camel to pass through them in the ordinary manner, or even if loaded. When a laden camel has to pass through one of these entrances, it kneels down, its load is removed, and then it shuffles through on its knees.' So, you see, that, although difficult for a camel to go through the eye of a needle, it is not impossible. If this is the meaning of our Lord's words, as some think, it is therefore not impossible for a rich man to enter into the kingdom of God if he will but humble

8

himself and cast off the burden of his riches."

"I suppose, then," said Clara, "that my verse can be explained too : 'Ye blind guides, which strain at a gnat, and swallow a camel.'" *

"I am not surprised that you should wonder at it, Clara," replied her governess, kindly, "but the expression refers to the custom of the Pharisees in straining all liquids before drinking them, lest they might accidentally swallow some insect which by the law of Moses the Jews were forbidden to eat because it was 'unclean.' Their actions were often so inconsistent with this fastidiousness or show of obedience to the law that our Lord rebuked them as straining *out* a gnat (as it is in the Revised Version), yet swallowing other things as large as a camel."

"Thank you, Miss Harson," said Malcolm. "I had it down too, for I didn't understand it any better than Clara did, but it seems very plain now, and I'm glad there's so much to learn about the camel."

* Matt. xxiii. 24.

"I think he's ever so nice," said little Edith, "and I don't believe he bites much."

Every one laughed at this charitable conclusion, and Miss Harson replied, with a loving caress,

"We will believe the best we can of him, dear. And it is really wonderful to see in how many ways this great awkward animal is useful.—How many things do you know already, Clara, that he can do?"

"Carrying people and things," was the reply, "and letters, and giving milk, and—"

But Clara could think of nothing more, and Malcolm asked,

"Do the people in the East ever eat camels' flesh, Miss Harson?"

"Yes," replied his governess, "the Arabs consider it a great delicacy and look forward to a camel-feast as a wonderful treat. All Eastern nations except the Jews—to whom it is forbidden as 'unclean'—eat camel-meat whenever they can get it; but travelers describe it as tough and stringy, with no particular taste. The hump is considered the choicest part, and this is always given to the most important guest."

"Is there anything else?" asked Clara.

"There are several things," was the reply. "You have heard of camels' hair, I think?"

"Yes," said the little girl, hesitatingly; "but does it come off camels?"

Miss Harson could not help smiling, while Malcolm laughed outright.

"You did not expect to find it on horses or elephants, did you?" asked the latter.

"Never mind, dear," said the governess, kindly; "Malcolm sometimes makes mistakes too. But camels' hair really does come off camels, and it is put to many valuable uses. For one of them I should like to have you find the sixth verse of the first chapter of Mark."

Clara read very thoughtfully:

"'And John was clothed with camel's hair, and with a girdle of a skin about his loins; and he did eat locusts and wild honey.'"

"The skin girdle, too, probably came from the camel; but the hair is very valuable. At the season of the year when it becomes quite loose it is pulled off in tufts; at other times the camels are shorn like sheep. The Arab women then spin it into strong thread, when it is ready for weaving. Many different fabrics are made of it, some of which are very coarse and rough, like the 'black tents' of the Bedouin Arabs, similar to those in which Abraham lived, and the rugs, carpets and cordage of wandering tribes. Mantles too are made of camel's hair, like the dress of John the Baptist, but they are usually very coarse. The best of the hair grows on the back and around the hump of the animal, as it is much longer there than on any other part. 'There is also a very little fine under-wool, which is carefully gathered; and when a sufficient quantity is procured, it is spun and woven into garments. Shawls of this material are even now as valuable as those which are made from the Cashmere goat.'"

"I wonder," said Edith, "if camels like to have their hair pulled out?"

Miss Harson did not think that part of it was considered, but she added:

"We must remember, Edie, that the hair is loose when it is pulled out, and the animals probably do not mind it; perhaps, too,

they are glad to get rid of the hair. A
dead camel is carefully skinned, and the
skin is made into a sort of leather which
furnishes its owners with sandals and leggins. Water-bottles, too, are sometimes
made from it, and it holds liquids better
than that of the goat. Indeed, this skin is
used in many different ways, and, with the
flesh and the hair, it makes the camel a
most valuable animal to the 'sons of the
desert.' Even camel-dung has its uses, as
it is composed chiefly of fragments of spicy
shrubs, and in the desert it is burned for
fuel. After being mixed with bits of straw
it is dried in the sun, and can then be kept
until it is needed for use. 'Mixed with clay
and straw, it is most valuable as a kind of
mortar or cement with which the walls of
huts are rendered weatherproof, and the
same material is used in the better-class
houses to make a sort of terrace on the
flat roof. This must be waterproof in
order to withstand the wet of the rainy
season, and no other material answers the
purpose so well. So strangely hard and
firm is this composition that stoves are

made of it. These stoves are made like jars, and have the faculty of resisting the power of the enclosed fire. Even after it is burned it has its uses, the ashes being employed in the manufacture of sal-ammoniac.'"

"Well," said Malcolm, "that must be a queer country to live in, and the camel's a queerer animal than I thought it was."

"Don't camels cost a great deal, Miss Harson?" asked Clara.

"No," was the reply; "nothing costs a great deal in that region, and a very good ordinary camel can be bought for a few dollars. The finest kind of dromedary is, of course, more expensive. And this reminds me of a camel-market in Tartary where immense numbers of these animals are bought and sold. 'In the centre of the town, it seems, there is a large square where the animals are ranged in long rows together, their front feet raised upon mud elevations constructed expressly for the purpose, the object of which is to show off the size and height of the ungainly creatures. The confusion and noise of

this market are said to be something fright-
ful and indescribable, with the continua.
chattering of the buyers and sellers dis-
puting noisily over their bargains, in addition to the wild shrieking of the camels,
whose noses are pulled roughly to make
them show off their agility in rising and
kneeling.'"

"Does a camel really shriek, Miss Harson?" asked Edith, who was more and
more surprised at the accounts of this
remarkable animal.

"Yes, Edie; travelers speak of its 'prolonged, piercing cry' as something very unpleasant, and 'the air of Blue Town,' where
the camel-market is, 'is made hideous with
the shrieking of the camels, as, to test their
strength, they are made to kneel while one
thing after another is piled on their backs,
and made to rise under each new burden
until they can rise no longer. Sometimes,
while the camel is kneeling, a man gets
upon its hind heels and holds on by the
long hair of its hump; if the camel can rise
then, it is considered an animal of superior
power.'"

"Well, I should think so!" exclaimed Malcolm. "But what cruel creatures those camel-men must be!"

"I am afraid that cruelty is not confined to them," was the reply. "And just look at that clock! Unless I send you all to bed at once, I shall deserve to be fined by the Society for the Prevention of Cruelty to Children."

CHAPTER VI.

MORE ABOUT CAMELS.

THE little Kyles were much like other children in never wanting to go to bed, no matter how sleepy they were, and it was very hard indeed to leave "those delightful camels," with which they were perfectly fascinated, and promptly obey their governess. But, besides their love for Miss Harson, they knew that unless they did obey there would be no more camels, and it was not half so interesting to read about them themselves.

Edith dreamed that night that a great camel was running after her and shrieking as loud as he could; but when Miss Harson came up and asked him if he were not ashamed of himself to tease a little girl, he looked very silly and scampered out of sight. They were all much amused at this dream, and Miss Harson said that Edie

A NOVEL RIDE.

must not dwell so much upon what was
disagreeable or alarming in their talks
about animals, because they were quite
safe from them, and these things could
not be avoided in order to understand how
these creatures differed from one another.
Miss Harson told them of a visit that she
had paid when she was quite a young girl
to a menagerie, where she saw several children riding upon a camel, and this pleased
the little Kyles very much.

"Now," continued the young lady, "I
wish to tell you of an important use to
which the camel is put, but it is one which
I hope you will never be obliged to try. I
have spoken of its singular stomach with
cells in it where water can be kept for a
long time, and 'this curious power of the
camel has often proved to be the salvation
of its owner. It has happened that when
travelers have been passing over the desert
their supply of water has been exhausted,
partly by the travelers and partly by the
burning heat, which causes it to evaporate
through the pores of the goatskin bottle
in which it was carried. Then the next

well, where they had intended to refill
their skins and refresh themselves, has
proved dry, and all the party has seemed
doomed to die of thirst. In these circumstances only one chance of escape is left
them: they kill a camel, and from its stomach they procure water enough to sustain
life for a little longer, and perhaps to
enable them to reach a well or a fountain
in which water still remains. The water
which is thus obtained is unaltered except
by a greenish hue, the result of mixing with
the remains of herbage in the cells. It is,
of course, very disagreeable, but those who
are dying from thirst cannot afford to be
fastidious, and to them the water is a most
delicious draught.'"

"Ugh!" said Clara, with *such* a disgusted
face; "I'd rather die of thirst than drink
that water."

"You may think so now," was the reply;
"but if the danger were really before you,
you would change your mind. The muddy
pools which are called 'fountains' in the
desert are much worse than this greenish
water from the camel's stomach, which, after

being kept for a few days, becomes quite fresh and clear. Another wonderful thing about the camel is that when it needs water it can scent it at such a distance that its rider has sometimes given up all hope of being saved from a dreadful death by thirst, when off would go the animal at full speed in a certain direction, until he stopped beside a desert spring that seemed to the fainting man a miracle from heaven. A sacred fountain at Mecca was first discovered by two thirsty camels."

"I thought that camels didn't need water as horses do?" said Malcolm.

"They do not need to drink water so often as horses and other animals do, because they can take in a supply that will last for some time; but they require a great deal of water, and they suffer as much as any other creature when the supply is exhausted. A thirsty camel is utterly unmanageable, and will not heed driver or rider when plunging madly after a fancied scent of water. But it is not often deceived; and when it has gained its object, great is the rejoicing among the exhausted travelers."

"Is the camel a wild animal, Miss Harson?" asked Malcolm. "I mean is it ever wild?"

"No," replied his governess; "it is peculiar in this respect. There are wild horses and asses and goats, even sheep and oxen, but a wild camel has never been found. Yet it is often the least tame and the most wicked of all domestic animals."

"Is it really wicked?" said Edith, in dismay.

"'Ugly' would perhaps describe it better. But it has shown itself to be revengeful as well as bad-tempered. Its stupidity is often mentioned, and it shows this in its habit of plodding straight on, no matter how it is loaded or what may happen to be in the way. 'As it passes through the narrow streets of an Oriental city, laden with goods that project on either side and nearly fill up the thoroughfare, it causes singular inconvenience, forcing every one who is in front of it to press himself closely to the wall and to make way for the enormous beast as it plods along. The driver or rider generally gives notice by continually calling to the pedestrians to get out of the way, but a laden camel rarely passes through a long street without having knocked down a man or two or driven before it a few riders on asses who cannot pass between the camel and the wall. One source of danger to its rider is to be found in the low archways which span so many of the streets. They are just high enough to permit a laden camel to pass under them, but are so low that they leave no room for a rider. The natives, who are accustomed to this style of architecture, are always ready for an archway; and when the rider sees one which will not allow him to retain his seat, he slips to the ground, and remounts on the other side of the obstacle.' But strangers are not apt to be so fortunate, and one traveler had a very narrow escape. After passing one or two of these archways by bending his head forward, he was talking to one of the party behind him, without thinking of what might be before him, when suddenly he noticed a shouting and running of the people in the street, and found, to his horror, that the camel was just beginning to pass through a gateway. It could not be stopped now, and its rider could only throw himself back as far as possible, with a feeling almost of certainty that he would never get through alive. His shirt-studs scraped upon the stone-work above, but he was mercifully preserved, almost breathless, yet alive. He says, though, 'If there had been a single projecting stone to stop my progress, the camel would have struggled to get free, and my chest must have been crushed in.'"

"O—h!" said the children, drawing a long breath of relief, as if they had just come from under an archway. "Wasn't that dreadful?"

"It certainly was, and it seems a pity that the camel has not sense enough to stop when it comes to such a place. But it appears very indifferent to its rider; and should the latter slip from its back in the desert, it will plod on the same as ever. It can seldom find its way home when it strays off anywhere, but it does not seem to care in the least whether it is with its old master or a new one."

9

"Doesn't the camel ever love any one?" asked Clara.

"Not very often, I think, but it has been known to hate people very vigorously. The camel is described as ill-tempered and revengeful, and there is a story told of one who had been unmercifully whipped by his driver. The man saw from the expression of the animal's eye that he deeply resented this treatment and only waited his opportunity for revenge; he therefore watched him constantly. One night, after retiring inside his tent, he left his striped cloak outside spread over the camel's wooden saddle; and this is what happened: 'During the night he heard the camel approach the object, and after satisfying himself, by smell or otherwise, that it was his master's cloak, and believing that the man was asleep beneath it, he lay down and rolled backward and forward over the cloak, evidently much gratified by the crackling and smashing of the saddle under his weight, and fully persuaded that the bones of his master were broken to pieces. After a time he rose, contemplated with great contentment the disordered mass, still covered by the cloak, and retired. Next morning, at the usual hour for loading, the master, who from the interior of his tent had heard this agreeable process going on, presented himself to the camel. The disappointed animal was in such a rage on seeing his master safe before him that he broke his heart and died on the spot.'

"This is an Arab story," added Miss Harson, " and it may not be true. Yet, from what I have read of this animal, I think that it might be. The cruel driver certainly deserved to lose his camel for treating him so barbarously, and he had a narrow escape in getting off with his own life."

"I wonder," said Malcolm, "if camels are so ill-natured and bad-tempered, that they ever let people lead them or do anything that they do not like."

"They would not, I suppose, if they were ever free," replied his governess, "but they are born in a state of submission, and they have never seen one of their kind going about as it chose. A traveler says that 'while being laden camels testify their dislike

to any packet which looks unsatisfactory
in point of size or weight as it is carried
past them, though when it is once on their
backs they continue to bear it with the patient expression of countenance which, I
fear, passes for more than it is worth.'"

"What funny creatures they are!" said
Clara. "Do their masters always treat them
as badly as that driver treated his camel,
Miss Harson?"

"Not always; sometimes they are very
kind to them, but this kindness cannot be
depended on. They will overload and over-
drive them, and yet they look upon them as
their most valuable possessions, often talking to them and encouraging them on their
journeys, and about the middle of the day
they will begin singing to them, and keep it
up for hours without stopping."

"What do they sing?" asked Edith.
"And do the camels like it?"

"They sing endless verses like this, dear,
and the camels are delighted, for they are
said to be 'greatly taken with music and
melody.' The camels' bells tinkle while the
Bedouin sings: ·

" ' Dear unto me as the sight of mine eyes
    Art thou, O my camel !
Precious to me as the health of my life
    Art thou, O my camel !
Sweet to my ears is the sound
    Of thy tinkling bells, O my camel !
And sweet to thy listening ears
    Is the sound of my evening-song.' "

"That is a very funny song," said Malcolm; "nothing seems to rhyme."

"It probably rhymes in Arabic," was the reply, "and probably, too, the camels are not particular about the rhyme, but they certainly seem to enjoy the singing."

The children thought this "very queer," but Miss Harson continued:

"A great many animals like music, and I have read lately of an unruly cow that allowed herself to be milked, and of a wild horse who would permit the groom to catch him by being drawn near the open parlor window where a lady played on the piano. The horse could be coaxed only by tender, sad things, while the cow liked martial music."

"What next?" thought the children, in amazement.

"It seems," said their governess, pres

ently, "that camels 'are fond of kneeling
at night just behind the ring of Arabs who
squat around the fire, and they stretch their
heads over their masters' shoulders to snuff
up the heat and the smoke, which seem greatly to content them.' Neither do they feed
entirely upon thorns, for an Eastern traveler encountered a respectable driver who
had driven three camels in the caravan, and
he says it was amusing to see his preparations for their evening's entertainment.
The tablecloth—a circular piece of leather
—was duly spread on the ground; on this
he poured the quantity of *dourrah* destined
for their meal, and, calling his camels, they
came and took each its place at the feast.
It is quaint to see how each in its turn eats,
so gravely and so quietly, stretching his
long neck into the middle of the heap, then
raising his head to masticate each mouthful
—all so slowly and with such gusto.'"

"I am glad they do get something to eat
sometimes," said Clara, "and I dare say
that if people would be good to them
they'd be very nice animals."

"Kindness will certainly do wonders,"

replied Miss Harson; "yet I doubt if we should ever think the camel 'very nice,' because he seems to have no power of loving. But I have been expecting some one to ask me a question."

"I will ask it, then," said Malcolm, promptly. "Are you going to give us a story, Miss Harson?"

"No," replied his governess, laughing; "that was not at all what I meant. Why has no one asked me what *dourrah* is?"

Edith hastened to say,

"Why, I meant to, Miss Harson, and then I forgot it."

"Then I will tell you without your asking, Edie. *Dourrah* is a species of grain that is cultivated by the Arabs, and with camels it takes the place of oats. I think, now, that we have studied the subject quite thoroughly, and it is time to look around for another beast of burden."

"Are not we to have a story?" asked Clara, in a beseeching tone.

"I do not know, dear, of any particular camel-story," said Miss Harson. "There are one or two fables in which this animal

figures, and the first is quite characteristic
—if not of the camel, at least of the camel-
driver.   An Arab had loaded his camel, as
usual, with as much as the poor beast could
well stagger under, and had—*not* as usual
—asked him whether he preferred going
up hill or down.   But the camel's reply
had nothing to do with hills. 'Pray, master,' said he, significantly, ' what has become
of the straight way across the plain?"

"Good for the camel!" said Malcolm,
approvingly.

"And what did the other one do?" asked
Edith, eagerly; for Miss Harson said "one
or *two* fables," and the little girl had no
idea of being put off with the smaller
number.

The governess laughed as she pinched
Edith's rosy cheek, and continued:

"Well, little girl, 'the other one' did
something not very different from the conduct of the donkey who wanted to be a
lap-dog.   He went, it seems, to an entertainment at which the various animals were
present, and a graceful monkey entertained
the company with his dancing.   He was

highly applauded, and received so much attention on account of his accomplishment that the camel, imagining that he could do it just as well, suddenly jumped up and began to dance too. But his awkward antics were so utterly absurd that the assembly burst out laughing, and ended by driving him out of the place. The moral of this is, Do not in public attempt a thing which you are not quite sure of being able to do well."

CHAPTER VII.

TO THE NORTH POLE.

"TO-DAY," said Miss Harson to her little flock, "we are going to take a wonderful jump in countries—from what may be called the region of perpetual heat, where our friend with the hump flourishes, to the region of almost perpetual cold near the north pole. What animal do we find here that is quite as useful to the people of cold countries as the camel is to the Eastern nations?"

Three young heads were busily thinking, and presently Malcolm said,

"Isn't it the reindeer, Miss Harson?"

"I think it must be," was the smiling reply, "and I have found a great deal that is interesting about this Arctic horse, and also about the curious people to whom he is so valuable. We shall certainly enjoy becoming acquainted with them."

138

The children were quite prepared in advance to be delighted; and when they saw the picture which their governess showed them of a reindeer, they studied it with great interest.

REINDEER.

"I think," said Clara, taking a side-view of the animal, "that it looks a little like a cow."

Malcolm seemed wonderfully amused at this idea, and Edith mildly suggested that "cows didn't have such queer horns."

'Well, they look just stuck on afterward," persisted Clara, "and I do think it's

face is like a cow's.—Isn't it, Miss Harson ?"

" It certainly has something of that look,"
was the reply, "and reindeer seem to me
like animals that are about two-thirds deer
and one-third cow. But in their activity
and restlessness they are all deer, yet they
are the only species of that wild family that
have ever been turned into domestic animals."

"Are the reindeer wild too ?" asked Malcolm.

" Yes ; they were all originally wild, but
many of them are now born in captivity,
as a well-to-do Laplander has large herds
of these animals. Wild ones are still
caught and trained, but neither the catching nor the training is at all easy. They
are caught with a lasso, as wild horses on
the plains are, and they object quite as
much to being caught. The lasso is often
thrown to a distance of thirty or forty feet,
and the would-be capturer is frequently himself thrown to the ground in the strug-
gle. As the reindeer runs, hoping to escape, the lasso draws tighter and tighter,

until at last he falls completely entangled
in the toils prepared for him. He is then
taken to his master's hut or farm, as the
case may be, but he is a very uneasy prize,
and it requires much patience and perse-
verance to train him and make him contented with his new quarters."

"I should think he'd break loose and run
away," said Malcolm, " where he could have
a much better time."

"That is the trouble, for he does not
want to work nor to feel that he cannot go
just where and when he pleases. Reindeer are allowed a great deal of liberty
and are seldom locked up in stables, as
they much prefer sleeping outside in the
cold. After a reindeer is caught it takes
a long time—about two years—to train
it to be of any use, and it must have a
lesson every day. While under training
the animals are fed with salt and angelica,
and are well treated in every way."

"Miss Harson," said Clara, remembering
the *dourrah*, "what is 'angelica'?"

"It is a plant," replied her governess,
smiling at her pupil's prompt question,

"the stalks of which are not unlike those of rhubarb. It appears to be used as a vegetable, and I have seen the stalks at the confectioner's in a candied state, like ginger, and of a delicate green color."

Anything that could be made into candy was especially interesting, and Miss Harson was eagerly asked how it tasted and if it were not possible for the youthful Elmridgers to find this out from actual experience.

"I remember once tasting a piece," said the young lady, "and I did not think it had much taste of any kind. But I will see that there is some angelica in the next box of candy that finds its way to Elmridge."

Boxes of candy did not find their way there very often, for neither papa nor Miss Harson thought candy desirable for a steady diet, but they were sure to appear at Christmastide and other extra seasons, and the children now felt quite satisfied with their prospects for knowing just how angelica tastes.

"Now," continued Miss Harson, "we will return to the reindeer, which we have rather neglected because he could not be made into candy, like angelica."

A merry burst of laughter rang out at the idea, and there seemed to be no getting back to the proper subject.

"'Two men,' says an Arctic traveler, 'came into camp with a young reindeer, and soon afterward the work of teaching him to draw a sleigh began. A long and very strong leather rein was attached to the base of his horns, and the rest of the harness was carefully attended to. The trace attached to the sleigh was several yards in length, the trainer himself being at quite a distance, thus placing the animal and the sleigh far apart. As soon as the reindeer was urged forward he plunged and kicked wildly, and it required all the strength of the man to hold him. After repeated rests for the animal and the driver the lesson was recommenced, and continued until the man became utterly exhausted.'"

"Does just one reindeer draw a sleigh?" asked Edith.

"Very often, dear," was the reply, "for the Arctic sleighs are quite different from

ours, and one of them does not appear too
much of a load for one such animal. These
vehicles are very light and only about six
feet long, and they have been described as
looking like half of an Indian canoe that
had been cut in two in the middle. The
sleigh has a keel like a boat; and the higher this is, the quicker it can travel through
the snow. There are various kinds of
these sleighs, some being used for carrying goods and provisions of all kinds, and
these have their tops covered, while others
are more elegant, being partly covered and
having cushioned backs. The latter can be
drawn rapidly, but all of them hold but one
person and are drawn by one reindeer."

"That is not nice a bit, "said Clara—"to
go out sleighriding alone."

"The Laplanders do not seem to agree
with you," said Miss Harson, "but I do not
suppose that they ever think of such a
thing as taking a sleighride merely for the
pleasure of it. Travelers say that it is very
difficult for those who are not used to them
to keep in these sleighs, and there is always
danger of being upset. 'You must make

up your mind,' says one who has tried it, 'to be upset a great many times before you learn to drive reindeer.' The most dangerous time is in going down a steep hill, as then the sleigh goes faster than the reindeer, fast as he is, and it is very hard to balance one's self and keep the vehicle from upsetting. The reindeer always quickens his pace in going down hill and carries his neck forward. 'I could hear all the time,' said the same traveler, 'a sound as if two pieces of wood were knocking against each other: this was produced by the feet. Every time the hoof touched the snow it spread open, and as it was raised the two sides were brought together again. Going down hill the pace was so rapid that the animals' feet seemed hardly to touch the snowy ground; they knew that if they did not go fast enough the *pulka* (sleigh) would strike against their legs.' "

"That is like coasting," said Malcolm; "it must be splendid. The boys in Lapland have fine times."

"I do not think we should consider them 'fine times' for any one. How would you

10

like, for instance, to live in a tent all the time?"

"Why, that would be fun, Miss Harson."

"For a month or so, perhaps; but suppose that you had a great herd of reindeer to watch, and that they were your only means of support, and you had to keep moving constantly in whatever direction they could find moss after they had eaten it all up in the last place? You would not find it very funny; and the inside of one of these tents is not a pleasant residence for those who are accustomed to nice houses."

"Our tent that papa had put up for us is nice," said Edith, "but I shouldn't want to live in it all the time."

"You could not have your cabinet of minerals in a tent, Malcolm," said Clara, who appeared to think there was danger that her brother might leave them all to go and live in a tent in Lapland.

Malcolm thought a great deal of his minerals, which he had collected with much care, and papa had presented him on his last birthday with a pretty cabinet to keep them in. The young gentleman's own room was quite luxurious, and not exactly like that of a person who would take kindly to living in a tent; besides his cabinet of minerals, he had a printing-press, a photograph-apparatus, and almost everything a boy of his age would be likely to enjoy.

"What do they live in tents for?" he asked, presently, referring to the Lapps, and not to the reindeer. "Should think they'd freeze."

"They live in tents," replied his governess, "because these light dwellings are more easily moved than houses. But I think I must tell you something about these curious people that you may understand how valuable the reindeer is to them. They are very short people, the tallest of the men measuring scarcely over five feet."

"Ho!" exclaimed Malcolm, scornfully; "I'm as tall as that."

"You would make quite an important-looking Lapp man," said Miss Harson, smiling, "and Clara would make a short Lapp woman. You see that these people are well suited to living in tents. They have short round faces and high cheekbones; men as well as women wear long hair."

"What funny-looking little things they must be!" said Clara. "What kind of clothes do they wear, Miss Harson?"

"In winter they dress much more warmly than we do, and a full suit is made of deer-skin, the coat reaching nearly to the knees and being fastened round the waist by a broad belt; the hair is left on these skins and turned inside for warmth. For shoes they make a pair of high moccasins from the skin which covers the legs of the deer, be-cause the hair of that portion is short and firm and wears better than any other part. These shoes are often ornamented, and they are generally worn very large; so that the Lapps can wrap their feet in a layer of dried grass, which is found in most of them."

"I should think it would feel horrid," said Malcolm. "We could not walk with dried grass in our shoes."

"It reminds me of a queer little girl I used to go to school with," continued his governess, "who often appeared in slippers

LAPLANDERS.

that were much too long for her and had the toes stuffed with cotton."

"Did she come from Lapland?" asked Edith.

"No, dear," was the laughing reply, "and, although she had cotton in the toes of her slippers, I am quite sure that she never had any dried grass. These Lapp men wear cloth and fur caps of a very peculiar shape, having a broad band around the forehead and a large square crown with small tassels fastened to the corners. The women are dressed very much like the men, and in summer both wear a long woolen garment, with leggins and shoes of reindeer leather."

The children thought this a hot dress for summer, but Miss Harson explained to them that summer in Lapland is a different thing from summer in the temperate zone.

"The Lapps do not have many suits of clothes at a time," continued the young lady, "as they are very seldom changed; so they do not need closets and bureaus, and other conveniences which take up room in a house. It is quite puzzling, though, how they manage to get all the family belongings, besides the people, into a tent, for these dwellings are very small. They look like Indian wigwams, and have a hole in the top to let out the smoke. Wood is scarce, and the fire is often made of juniper-branches."

"What do they make the tents of?" asked Malcolm. "I suppose reindeer-skin is what they use."

"Not right this time, Malcolm, although I should probably have said 'reindeer-skin' myself had I not been reading a description of these tents, which you shall hear:"

"'The tent used by the Laplanders is portable, and is conveyed from place to place by the reindeer. Its frame is composed of poles fitting into one another, easily put together, and so strong and well knit that they can resist the pressure of the heaviest storm: a cross-pole, high up, sustains an iron chain, at the end of which is a hook to hold the kettles. Over the frame is drawn a cloth of coarse wool, called *bad-mal*, made by themselves, no skins being

ever used; it is composed of two pieces,
and is made fast by strings and pins and
well secured. The porous quality of the
cloth permits a partial circulation of air. A
small door made of canvas is suspended at
the top of the entrance. The woolen cloth
is exceedingly durable, often lasting more
than twenty years. The tents are frequently much patched, for a new covering
costs from thirty to forty dollars. In summer these tents are usually pitched near a
spring or stream of water where the dwarf-
birch and juniper furnish fuel, and not far
distant from good pasture. This is called an
encampment, and it is the kind of house in
which live the mountain or nomad Lapps,
who wander with their reindeer from pasture to pasture all the year round. The
sea Lapps, who live along the wild coasts
and get their living by codfishing and acting as sailors to the Norwegians, live in
queer-looking mounds of earth with holes
in the top. Better dwellings are shaped
like houses and built of turf, having sometimes walls of stone to protect the turf and
make the houses stronger. Some which

are considered to be very grand residences are built of logs.

" The forest Lapps, who live in the woods, seem to fancy very queer-shaped houses. In some districts the lower part is square, built of three or four logs joined together, the upper portion being pyramidal, of split trees covered with birch bark, over which boards are put. A large flat stone is in the middle for fire, at the place where the cooking is done. There is the usual hole in the centre for the escape of smoke. In some houses the floor is covered with stone slabs, in others with young branches of birch, and, as in the tent, skins are spread for the family to sleep upon. Near the dwellings are large enclosures where at a certain season of the year the reindeer are penned every day. Many of the forest Lapps own extensive herds. The farming Lapps, whose dwellings resemble those of the forest Lapps, generally live together in small hamlets on the shores of rivers or lakes. A church, a parsonage and a schoolhouse are always to be seen in one of these settlements, and many of the Lapps are very

religious people.   They are always kind to
strangers, and will take them in and give
them food and shelter at any hour of the
night, although wakened from a sound
sleep to do so."

"Then they are not heathens at all?" exclaimed Clara, in surprise.

"No, dear," replied her governess; "they
are generally very good little people, and
are honest, peaceable and generous.   Many
of their customs would seem strange to us,
and not altogether pleasant.   A traveler
among them says that 'the rich man lives
in the like smoky and filthy hut as the
poor man, only it is larger, because it must
be, to accommodate his larger family; for
his servants or herders are strictly members
of his family and live on an apparent equal-
ity with himself.   The great kettle is hung
over the fire in the middle of the hut and
filled with the flesh of the reindeer; and
when it is boiled, all go and help themselves alike, with fingers or sticks or with
forks and spoons made of the bones or
antlers of the deer, or their sheath-knives,
which always hang at the hip of young and

LAPLANDER AND HIS FAMILY.

old. All sleep together in the hut on the pallets of deerskins wherever they can find room.' "

Clara and Edith were quite sure that they didn't wish to go among the Lapps, but Malcolm still thought it would be "fun."

" The only difference between rich and poor seems to be that the wealthy man will own many thousand reindeer, while a poor man may have but two or three. Many things are paid for in this rather clumsy coin, and a good servant receives from three to six reindeer for a year's work. The Lapp women are industrious and make all the clothing for the family. This must be hard work, as skins are not easy to sew, and the only thread they have is made from deer-sinews. They weave all the cloth, and also embroider on cloth and leather. The men, too, are constantly busy watching the reindeer, who will scatter in all directions if they are frightened, and with wolves always watching, too, it is necessary to be continually on the lookout. 'If the wolves are not hungry, they will not dare to come near; but if in want of food, they will at-

tack a herd in spite of all precautions.
Often the deer by the sense of smell detect
the approach of their enemies; in that case
the herd moves away. The Laplanders
then know what to expect, and with their
dogs pursue the wolves, keeping the deer
together at the same time.'"

The children were all surprised to hear
of any dogs in Lapland.

"What kind of dogs are they, Miss Harson?" asked Malcolm. "Do they look like
ours?"

"They are not large," was the reply,
" and they do not look much like our dogs.
They have long, thick hair, and some of
them resemble small bears, being of a
dark-brown color and without tails. 'It is
wonderful to see how these dogs can keep
a flock of reindeer together; occasionally,
for some unknown reason, a panic seizes a
herd, and it takes all their cunning and a
great deal of running to prevent the deer
from scattering in all directions. These
dogs are the useful friends of the Laplanders. In order to keep them hardy, strong
and healthy, they are treated roughly,

are never overfed, and are not allowed to rest till their owner does; indeed, they often seem to get only the food they can steal.   Every man, woman, grown child and maid-servant has his or her own dogs, which obey and listen only to the voice of their owner.   They are exceedingly brave, and not afraid of wolves and bears, which they attack without fear, but with great cunning, taking care not to be bitten by them and choosing their time and place to bite.' "

CHAPTER VIII.

REINDEER WAYS.

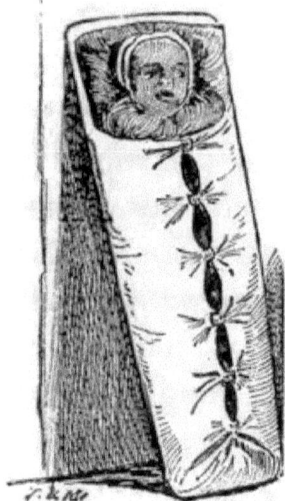

A QUEER CRADLE.

"MISS HARSON," asked Edith, "are there any little babies in Lapland?"

"Why, yes, dear," replied her governess, laughing; "there are babies wherever there are people, and the Lapp babies are very funny. The cradle, which is hung up, looks much like a big slipper; and when the baby is laid in it, his clothes are all taken off and he is covered with a sheet, a piece of coarse cloth and a sheepskin quilt. There are holes on each side of the cradle, and through these a stout cord is laced across to keep the

covers from falling off and the child from
tumbling out."

This did not sound like a pleasant way
of sleeping, but a Lapp baby, Miss Harson
said, never knows anything better.

"When the Lapps are traveling from
place to place," continued the young lady,
"these cradles—which are about two and a
half feet long and half a yard wide—are
slung on the mothers' shoulders and go
through some dreadful storms and snow-
drifts. The baby-reindeer are always carried, too, or put in a sleigh; and when
their mother calls them, they give a queer
kind of grunt, which the young ones answer."

The children thought it funny to carry
little animals or take them sleighriding, and
they had a great many questions to ask
about the baby-reindeer.

"Their color," replied Miss Harson, "is
at first nearly white, but it gets darker
until they are fully grown. The old ones
are gray, with coarse, thick hair nearly two
inches long, but it is longer and thicker in
winter than in summer. This hair is much

darker on the back and almost white underneath. The reindeer's horns are branching, like great pieces of seaweed, and the reindeer itself is a clumsy-looking animal with stout legs and very broad hoofs. 'The animals are never housed, for they like cold weather and snow. Food is never given them, and unless brought up to do so they will not touch the moss that has been gathered. They often will not even raise their heads as you approach them, and remain quiet when the Lapps pitch their tents. Some years prove unfavorable to their increase, on account of the amount of snow, which prevents them from digging for food; the herd then becomes weak and emaciated, and many die. The spring also is a bad time for them; the snow melts during the day and a thick crust forms at night, so that their feet break through, causing lameness and disease. The horns of the males, which often weigh forty pounds, attain their full size at the age of five or six years; those of the cow, at about four years. After the age of eight years the branches gradually drop off.

11

The shoulder-blades appear a little high, occasioning a slight hump or protuberance. Without the reindeer the Laplander could not exist in those northern regions; it is his horse, his beast of burden, his food, his clothing, his shoes and his gloves.'"

"It is queer about horns," said Clara. "I can't think how they grow. What makes 'em, Miss Harson?"

"They certainly are singular ornaments," replied her governess, "as they do not take the place of ears, and seem to be of no use whatever except as weapons of defence. What is still more strange about the reindeer, the does, or she-deer, have antlers as well as the stags, although of smaller size, but the does of other deer are never seen with them. We have already become acquainted with various kinds of horned animals, but deer-horns are entirely different from any others. The horns of the cow, the ox, the goat, etc. are hollow and composed of a substance like that of hoofs and talons; they never branch like those of the deer, and, once grown on the head of the animal, they stay there unless broken off by accident. But a deer has his horns only part of the year: they actually drop off like the leaves from the trees at the beginning of winter, and he does not get a new pair until spring."

"Why, that is like the crabs and lobsters," said little Edith, who was quite proud of remembering that these creatures changed their old shells for new ones.

"Yes, dear; they are like them in casting off a portion of their bodies and having it grow again. I will read you a description of the manner in which these antlers grow in the common deer, and, although there are some large words in it, I think you can understand how these strange horns are formed: 'We will suppose that a full-grown stag is hiding in the depths of the forests in the month of March. He has no horns of any kind, and is hardly to be distinguished from a doe but for his superior size. On his head are two slight prominences covered with a kind of velvety skin. In a few days the prominences become much larger, and in a week or so begin to assume a hornlike shape. Now grasp these

budding horns with your hand, and you will
find them quite hot—considerably hotter
than those of the young ox. They are hot
because this velvety substance with which
they are covered is little else than a thick
mass of arteries and veins through which
the blood is pouring almost with the rapidity of inflammation, depositing with every
touch a minute portion of horny matter.
More and more rapidly increases the
growth. The external carotid arteries become enlarged to supply a sufficient tide
of blood to the horns through their arteries,
whose size can be imagined from the grooves
that they leave on the horn. At this period
of their growth the horns can easily be broken off; and if they are wounded in any way,
the blood pours out with astonishing rapidity. At length the process is complete, and
the noble animal walks decorated proudly
with his enormous mass of horns. But the
horns are at present useless, or worse than
useless, to him, for not only does he not use
them, but he fears the slightest touch, because the sanguineous tide still pours round
them. How is this to be stopped? and

how is the velvety covering to be got rid of?'"

"Oh, Miss Harson!" exclaimed Malcolm, in a sort of comical despair; "what *is* 'sanguineous tide'?"

"Try and think it out for yourself," was the reply. "How did the *Sanguinaria*, or bloodroot, get its name?"

"I see now—thank you. It means the blood in the veins. But why couldn't the person who wrote it say so?"

"Perhaps because 'sanguineous tide' sounds better than 'blood' and makes more variety of expression, and the book was written for grown-up people. But now let us see how the deer gets rid of the velvet covering on his horns: 'In a manner no less simple than wonderful. The arteries, having completed their work in depositing sufficient matter for the substance of the horn, now turn their attention to the base. It will be seen that all the arteries that supply blood to the horns must necessarily pass up their base. As the bony substance is deposited each artery leaves for itself a groove very deep at the

base and becoming shallower toward the
tip. The entire horn being furnished, the
base now becomes enlarged ; the grooves
in which the arteries lie are covered by a
bony deposit that compresses the artery
within; the deposit becomes gradually thicker, and the arteries are in consequence
gradually reduced in size, until at last they
are completely obliterated and the supply
of blood is cut off entirely. The velvet,
being thus deprived of its nutriment, soon
dies, and in a few days dries up, when the
deer rubs off the shriveled fragments against
the trees and is ready for combat.' "

"How does he know just when to do it?"
asked Clara.

"I suppose, dear, that he is guided by feeling as well as by instinct; but he is sometimes too impatient to wait until the arteries
are quite dry, and rubs the velvet off while
the blood trickles down his horns. This
gives him quite a horrid appearance, and I
should think the animal would be apt to
feel very uncomfortable. The reindeer, it
seems, is quite disposed to hurry matters
in this way."

"The horns look queer and uneven in the pictures," said Malcolm—"more like a great bunch of thorns, or something, stuck on their heads. Aren't they very long, Miss Harson?"

"Yes; the measurement of the longest part is over a yard, and they have several spikes and branches."

"I should think they would get dreadfully tangled in trees and bushes when there are any," said Clara.

"No," replied her governess, "for in that case the antlers are laid back flat on the head, and really help the animals on their way instead of hindering them."

"Do reindeer go fast?" asked Malcolm. "As fast as horses?"

"Yes—even faster; but their speed depends upon the weather and the ground. They do best in autumn and early winter, because the cold agrees with them, and early in the season they do not get so tired digging for food as when the snow is deep. When the snow is well packed and furrows have been made by other sleighs, a swift reindeer will go a hundred and fifty miles in one day, and he can travel for five or six hours without stopping at all."

"Miss Harson," said Edith, with great interest, "will you please tell us what kind of things the reindeer finds to eat under the snow?"

"Yes, Edie. He finds the moss which he thinks the most delicious thing in the world, but which he is not willing to eat unless he digs it out for himself. A traveler in Lapland speaks of driving through a forest and coming suddenly upon a great number of reindeer who were in comical positions, having thrust their heads and horns and fore feet far down into the snow. They were busily digging for moss, first with one fore foot and then with the other; and as the holes grew larger still less of the reindeers' bodies would be seen, and the short, funny tail stuck up in a curious fashion. After going on for a while the traveler and his companion turned back, and drove again through the same forest. Then, as he says, 'another strange sight presented itself. Where had the reindeer gone? None were to be seen. Had they been taken away? As I approached the herd I discovered that all of them had dug holes so deep that I could see only their tails, which swayed to and fro. This was certainly a landscape I had never seen before.' "

It seemed very hard to the little audiencě that so comical a sight was thousands of miles away, and that they were not likely ever to see it, the waving of all those little tails above the snow was so extremely funny.

"Don't deer ever have any better tails than that?" asked Malcolm, quite contemptuously, as he examined the picture.

"No," replied his governess, laughing; "the animal seems to come to an end very suddenly. His tail is short even for a deer, but it is thicker than that of some other varieties. His head is certainly the handsomest part of him. But, handsome or not, he is a wonderfully useful animal, and seems to have been made especially for the people of the cold regions where he lives. Reindeer's milk is an important article of food among the Lapps, as it is

made into butter and cheese, while the
cream is also dried in empty bladders, and
mixed afterward, with hot water, into a
kind of porridge which travelers pronounce
good to the taste as well as nourishing."

"How funny," said Clara, "for deers to
give milk! Is it like our cows' milk, Miss
Harson?"

"It is better than that in one way, because it is much thicker and richer—more
like cream; but there is very little of it.
One deer will give scarcely enough at once
to fill a small cup, but it is too rich to drink
until water has been added to it. Milking-
time, too, is different from what it is with
us, as the reindeer objects to being milked
at all, and a lasso is sometimes thrown
over its horns to prevent it from running
away. Sometimes one woman holds the
deer while another does the milking."

"Well," said Malcolm, with a sigh of
regret, "I think Lapland's a splendid place,
and I wish I could have just one good
sleighride with a reindeer."

"Perhaps," replied his governess, smiling, "it will answer quite as well for me to

read you part of an account of 'one good sleighride with a reindeer' which a traveler in those cold regions took."

The children were all eager to hear of it, and the young lady continued:

"This sleighride lasted a whole day, and fortunately the traveler was not alone. The party were overtaken by a heavy snow-storm, with a furious wind, and were obliged to come to a sudden stop: 'We could go no farther, for it was impossible to see anything ahead, and there was danger of mistaking the passes which were to lead us to Norway. Besides, our reindeer needed rest, and from excessive thirst they were eating the snow ravenously. I shall never forget how the storm raged as we lay by a rock with our backs to the wind. For three hours we remained still, frequently almost buried, the thermometer being at fifteen degrees below zero. The wind was so terrific at times that hardly a particle of the several feet of snow that had fallen during the winter months remained on the ground. It flew in dense bodies, carried hither and thither. A hill was no sooner formed than it scattered in thick, heavy masses; we were fearful of being buried under one of these hillocks, which were as dangerous as those formed by sand in the desert of Sahara. I noticed by the quickening steps of my animal that we were approaching the slope of a hill. I was not mistaken, and we descended a long, steep declivity with fearful speed. Suddenly my reindeer sunk above his flank into a bank of unpacked snow, and before he had time to spring out my sleigh dashed quickly ahead of him, and, suddenly stopping, threw me out. Fortunately, I leaped in at once, and the animal again started at what I thought a greater speed than before. One of the Finlanders just in front of me was less fortunate: his sleigh, moving faster than his deer, struck upon the legs of the animal, and he was thrown out. I saw the danger at a glance, but, unable to stop, went rushing down in the same track. My sleigh struck his, and by the force of the collision I was pitched headforemost into the snow. To add to the confusion, my animal became mad and charged upon me; but I was soon on my legs and in again, following the Finn, who had started once more and was going at a rapid rate toward the base of the hill. Then came Elsa's turn to be upset, but soon she recovered her seat, and we reached the bottom without further mishap. The adventure was exciting and glorious.' "

"There!" exclaimed Malcolm, just as Miss Harson had thought he would. "Doesn't the traveler say it was glorious? It was just fun!"

"Wait a minute, my excitable young friend, and you will probably change your mind. The party have not yet reached their journey's end, and this is what came next: 'At the foot of the hill the snow thinly covered the frozen stream, and the scene became rather ludicrous. There was not snow enough to prevent the reindeers' hoofs from touching the ice, so it was an impossibility for them to advance a step. The awkward attempts they made were quite amusing. We were compelled to get out of the sleighs and lead the animals, and it was with considerable difficulty and with

great loss of time that we succeeded in crossing.   It is impossible for reindeer to travel
over ice.' "

"How strange that seems!" said Clara.
"I should think they'd have to travel over
ice when there's so much of it around."

"There is more hard snow than ice, for
ice is formed by partial melting.—' As we ascended the mountains on the other side the
snow became deeper; a part of the way
led us through very narrow ravines in
which it was so deep and soft that our
boatlike sleighs ploughed heavily through
it, sinking sometimes into it above their
sides.   I could not but admire the adapta-
tion of the reindeer for such traveling; their
hoofs, between which grows long hair,
spread in the snow as soon as their feet
touched it, and, although the depth must
have been in places eight or ten feet, they
seldom sank into it as deeply as their knees.
They moved so quickly that there was no
time for them to sink deeper.   At times,
however, when passing through a very soft
and heavy snowdrift, they would sink even
to their bellies.

"'Our progress now was exceedingly tedious. In ascending the hills our reindeer became very tired from their struggles in the snow. They were heated; their mouths would open and they panted for breath, sometimes even protruding their tongues. They were often so exhausted that they would drop upon the snow and lie on their backs, apparently in great suffering, then breathe hard and be so utterly helpless that a stranger would think they were about to die. After resting a few minutes in that position they would regain their breath, rise to their feet, eat snow and set off again. There were many steep and short hills up which it was impossible for them to run, and we were often obliged to get out of our sleighs to let them rest.

"'We came to the worst part of the journey on the brink of a narrow ravine, and stopped, for the descent was very abrupt and preparations to ensure safety had to be made. I felt rather concerned when I saw the difficulties to be encountered on the route, which was somewhat crooked; in some places the ridge over which we were to drive was quite narrow, the gneiss rocks were bare and the track was very steep and dangerous. While resting I watched the weary reindeer eating snow as fast as they could.

"'After every one had arrived the preparations began. Numbers of sleighs were lashed together by a long and strong leather-plaited cord, which was first secured to the forward part of each, then, passing along the middle, was made fast, after which it was attached to the next in the same manner, and so on; four others were connected with mine. In this way eight or ten were often fastened together. With the exception of the leader, each reindeer was secured to the rear of his sleigh by a leather cord from the base of the horns; almost every sleigh had a deer behind. Each man remained in his vehicle, the distance apart being small. The spare reindeer were for the first time harnessed, and the tired ones put behind.

"'Pehr had to start the whole train, which, when once put in motion, would go with great velocity; he rode with his legs outside, turned back somewhat, with his feet touching the snow. Every man but me seated himself in the same posture, the feet acting as rudder and drag in the snow. I was not allowed to ride in that way, for they said my legs would surely be broken. When everything was ready, Pehr looked back and gave the signal, and started his reindeer down the hill in a zigzag course. This required great dexterity, as we flew over the snow with astonishing speed. At times the sleighs would swerve on the declivity, but we went so fast that we were soon out of danger.

"'I was anxious in the highest degree: if one of those cords had broken, we should have been precipitated far below or dashed against the rocky sides. I admired the simplicity of the arrangements, which were dictated by the fact that reindeer cannot bear to be pulled by the head, especially by the horns; each one, therefore, makes an effort to disengage himself, and by so doing acts as a brake to the ones in front, so that no sleigh is likely to be overturned. But what a speed! what a precipice on our right! In

12

two or three places we went for a short distance over the bare rock. I was afraid the reindeer would miss their foothold, and was intensely excited, for I might at any moment have been thrown out headlong. Pehr and my other companions were accustomed to this route, and knew what they were about. After reaching the bottom of the ravine we allowed the panting animals to rest. We were now on the western shed of the mountains, and had just ended the most thrilling ride I had ever taken.'"

"The poor reindeer!" said Edith, pityingly, when her governess had finished. "How tired they must have been!"

"And the poor men with their feet on the snow!" said Clara. "How cold they must have been!"

"I suppose he was glad when he got down there," said Malcolm, thinking of the traveler, "but what a splendid time he had, though!"

Miss Harson was much amused by these remarks, and she said to her little flock,

"I think that such dangerous adventures are better to read about than to take part in, and I have read you this long description that you might understand just how the reindeer is used in sleighing. The account is taken from a very interesting book called *The Land of the Midnight Sun*, which you will each read for yourself after a while. We are indebted to this book for a great deal of information about the reindeer and the Lapps, and I think we shall not easily forget what we have learned."

"Miss Harson," asked Edith, rather timidly, "was that long sleighride meant for a story?"

"It isn't possible," was the reply, "that you would expect any other?"

The children did not look in the least alarmed even when their governess warned them that some day the machine would have to stop grinding; and when they discovered that it was to be one of Miss Harson's own stories, they were in a state of delighted expectation. The story was called

WHY CARL WENT BACK.

Little Carl Nosser was wakened from a sound sleep one winter night by such a

commotion that at first he thought a wolf
or a bear, or perhaps one or two of the
reindeer, had got into the tent; for Carl
lived in a tent in Lapland, and he did not
know of people who lived in any other
way. He sometimes had heard of travelers who roamed about the country and
asked a great many questions, but he never expected to see one of them, and he
troubled himself little about them. He
had plenty of work to do, although he was
only ten years old; and when night came,
he fell asleep as soon as he got under his
sheepskin.

Yes, Carl slept under a sheepskin, and
over one too, and in cold weather—that is,
cold weather for Lapland—he got into a
sort of bag made of reindeer-skin with the
fur inside. This was in place of a night-
gown, and it kept his toes—and indeed
every part of his whole person—as warm
as toast.

Carl liked to be warm, and he dearly liked
to be comfortable. It did not make him
comfortable to hear this noise and talking
in the middle of the night, and to think that

some dreadful creature must have got into the tent, and at first he tried to go to sleep again and forget all about it. But presently the thought of his mother and little Marta, and of his father, the kind, generous Lars, made him feel ashamed, and he started up, quite awake, to see what was the trouble.

It was a queer-looking place inside of the tent, only it did not look queer to Carl, because he was used to it. Right in the middle, under the opening at the top, a bright fire was burning, and over the fire was swung a large brass kettle with rein-deer-meat cooking in it. How good that meat did smell! and how Carl wished he could have some! They were making coffee, too, and he liked the smell of coffee, but they did not have this every day, and Carl had had his supper and been put to bed long ago. There were so many things in the tent that it seemed wonderful how room could be found for them all. There were only a large chest and some reindeer-skins on the floor to sit on, and the fire took up a great deal of space ; but there were several people and two or three dogs and various piles of wood, while hanging over-head were the baby's cradle and some rein-deer-horns, several pairs of snow-shoes, skins spread out to dry, harnesses and pieces of frozen meat. Besides all these things, clothes, saddles, empty pails, iron pots and wooden vessels were scattered around.

Carl had seen this confusion ever since he could remember, and he supposed that every one's home looked like this; but he had not seen the tall stranger who, with his guide, was now standing before the fire. The dogs had made such a terrible noise outside on the arrival of the visitors that it had wakened the boy, and now, in his comical sleeping-bag, he looked as strange to the English gentleman as the gentleman did to him. The other man was a Lapp, and Carl had seen him before, but he could not tell why he wanted to come and visit them in the middle of the night. He had come on very urgent business, for he and the gentleman were on their way to Sweden, but had met with an accident by which they lost one of their reindeer and nearly lost their lives, and, as honest Lars Nosser's encampment was the nearest shelter, they had ventured to stop there in the middle of the night and rouse the worthy people from their slumbers. A bed and something to eat were necessities that must be provided at once, and the kind hosts went about supplying them as cheerfully as if it had been the middle of the day.

After a while Carl's mother saw her boy's great round eyes wide open; and when the guests had been attended to, she slipped over to the bed with a nice little piece of reindeer-meat, saying, as she popped it into his mouth,

"Lie down to sleep again, little one; the stranger will not hurt thee."

"Then he isn't a giant, mother?" whispered Carl, who had heard of such beings.

"No, indeed!" was the reply; "he is a good man, and very tired. See! I am going to make his bed for him right away."

For a mattress Mrs. Lars spread fresh skins over a pile of young birch-tree branches in one corner of the tent, and, with other skins for comfortables, the unex-

pected visitors had a warm resting-place,
and were soon asleep. Carl had scarcely
swallowed his piece of meat before he forgot about that and everything else, and he
did not wake even when he was moved
out of the middle of the bed to make room
for the rest of the family.

The next morning the little Lapp boy
could recollect nothing of what had happened in the night before, and he was surprised all over again when he saw the
visitor. But he soon discovered that the
gentleman had pleasant, laughing eyes and
a ready smile, and this made him think that
one of the very first things he would want
to do would be to see his reindeer; for
Carl had a fine large reindeer of his own
which was never used without his permis-
sion, for it had been given to him as soon
as he was born—which is the Lapland fashion of doing with babies—and it seemed
to him that no other animal in the whole
herd could compare with his. Mr. Thorne
praised its handsome antlers and its pretty
color, and laughingly asked Carl what he
would take for it; but the little Lapp only

clasped his arms tightly around his reindeer's neck, and the animal replied with an
affectionate grunt.

Then Carl put on his snow-shoes, and
queer-looking things they were, like long
rods with turned-up ends. They were
made of fir-wood, and were four or five
inches wide and only about a third of an
inch thick in the middle, which was the
thickest part. There was a piece of birch
in the middle and over this a loop, through
which the boy passed his foot. The under
part was smooth and had a narrow furrow.
The length of these shoes was enormous
—fully equal to Carl's height—and to the
stranger they looked awkward and trouble-
some, but the young Lapp ran and jumped
with these strange appendages, and even
took flying leaps from one snowbank to
another, laughing gleefully over the excitement. It looked so easy that Mr. Thorne
tried it and attempted to slide down an
embankment, but he reached the bottom
sooner than he expected, and his shoes
came off as soon as he started. He too
laughed merrily, but Carl was quite grave

and polite, and seemed afraid that the
gentleman's feelings had been hurt.

The visitor stayed that day and night,
waiting for the guide to make some necessary preparations for their journey, and he
and Carl were constant companions. He
liked all the kind, hospitable family, but he
had taken a decided fancy to the bright
little fellow who seemed so partial to
his society, and he told him a great deal
about the wonderful things to be seen in
England. Carl listened eagerly; and when
Mr. Thorne asked him if he would like
to go home with him and go to school, he
said that he would.

The good Lars and Margarita looked
troubled at first when the gentleman proposed taking their boy so far away from
them, but after a few moments they said
reverently,

"God will take care of him as well there
as here. It is for Carl's good. We consent."

Then Carl got his beloved reindeer and
his snow-shoes ready, never doubting but
that he should need them in England; his

mothei made up a little parcel of clothing
for him, which she gave him with a good-
bye kiss, and his father solemnly blessed
him; so, feeling very important and quite
like a man, he started with his new friend.
After a while he began to ask.

"Do they have such beautiful snow as
this in England? Are there forests there
of birch and pine and fir? Do you have
warm skins to sleep under and nice blad-
der-puddings to eat? And do you hunt
wolves and bears?"

"Oh no," said Mr. Thorne; "our country is quite different from that. It is not
cold, and there are no wild animals. You
will not need your snow-shoes, and I am
afraid the reindeer will starve unless you
can get him to eat something else instead
of moss."

Carl looked solemnly at his companion
while he was speaking; and when he had
finished, he replied:

"Then I will go back to the mother and
stay in Lapland. It is not good to be
where everything is so strange;" and, in
spite of all his new friend's persuasion, the

little Lapp turned back on his snow-shoes, and never stopped until he found himself in front of the beloved tent with its nice hole in the top for letting out the smoke, its piles of wood and its sleigh, and all sorts of handy things, scattered about outside.

Mr. Thorne could not help laughing at Carl's quick movement toward home as soon as he realized that all the delightful things to which he was accustomed would have to be given up, but, in spite of his disappointment, he felt very kindly toward the little fellow, and when the guide returned sent him a box that was a constant source of delight as long as its contents lasted. On top of this box was a dear little Bible with Carl's name and "A gift from his English friend" written in it, while inside the box were candies the like of which the boy had never dreamed of, and just in the middle he came to a white-sugar reindeer. It was beautiful, and Mr. Thorne had tied to it a slip of paper on which was written "Selma," the name of Carl's own live reindeer.

"It must have come all the way from Stockholm," said Carl, in a tone of awe.

The little Bible, too, was a great prize, for, although the blessed book was well known and read aloud every day in the tent of Lars Nosser, Carl had not yet had one of "his very own."

The good Margarita clasped her boy in her arms and said,

"It is well: the home and the Bible are all any one needs."

CHAPTER IX

HAVING warmly approved of the story about the little Lapp boy, the children eagerly asked if there were any reindeer in the United States.

"Not any that are known by that name," replied Miss Harson, "but there is a species of deer called the woodland caribou which is thought by naturalists to be very much the same. The antlers are almost exactly alike, but the woodland caribou is often a third larger than the reindeer. The American species is also lighter in color and has more white about the neck, the old males having a long white mane which often measures over a foot in length. This mane curves very gracefully in front and is quite ornamental. The head and the legs of the caribou are of a tawny brown, while the rest of the body is nearly all white."

"But they can't find reindeer-moss to eat in this country, can they?" asked Malcolm.

"Yes," replied his governess; "a moss which answers this purpose is found in great quantities in Nova Scotia and New Brunswick, where the caribou is seen; and, besides this, it feeds upon a variety of lichens and grasses, also upon shrubs and trees. It is said that after they have disappeared an experienced hunter can easily follow these animals 'by noticing where they have cropped the twigs or stripped the moss from the trees in passing, and by careful inspection will judge something of their number and of how recently they have passed. This cropping is done by the animals without their stopping to feed, but as they walk along.' The moss of which they are so fond sometimes grows two or three feet deep in barren places where no other vegetation is found. It is very thick in Labrador, where the inhabitants almost entirely depend on the

caribou for their food.   A traveler in Nova
Scotia speaks of seeing the snow quite
trodden down during the night by this animal, 'which had come to the place to feed on
the "old men's beards" in the tops of the
spruces felled by the lumberers on the day
previous.   In the same locality,' he adds,
'I have observed such frequent scratchings
in the first light snows of the season at the
foot of the trees in beech-groves that I am
convinced that the animal, like the bear, is
partial to the rich food afforded by the
moss.'"

"Miss Harson," said Clara, "I remember you telling us a funny thing about the
reindeer: you said that its feet spread
open when it was traveling over the snow.
Does the caribou have the same queer
kind of feet?"

"The very same, as the hoof is large
and broad and seems made for bearing the
animal up in snow and on soft, swampy
ground.   The cleft between the toes is
long, and this enables the foot so to spread
out that sometimes it looks twice as large
as at others, 'and the imprint in soft

ground is so much larger than on a hard
surface as to require the eye of a practiced
hunter to recognize the track as made by
the same animal.' This wonderful foot
changes in the winter: the edges grow out
in thin, sharp ridges; the under part
hardens; and the caribou is provided with
a natural pair—or, rather, two pairs—of
skates which never get out of order.
'With this singular conformation of the
foot, its great lateral spread and the addi-
tional assistance afforded in maintaining a
foothold on slippery surfaces by the long,
stiff bristles which grow downward from
the fetlock, curving upward underneath between the divisions, the caribou is enabled
to proceed over crusted snow, to cross
frozen lakes or to ascend icy precipices
with an ease which places him, when in
flight, beyond the reach of all enemies
except, perhaps, the nimble and untiring
wolf.'"

A very strong desire was expressed by
the youthful audience to see the caribou
with his winter skates on, and Edith suggested that perhaps there might be some

13

in the woods at Elmridge when it was *very*
cold and slippery.

"No, dear," replied her governess;
"these animals are great wanderers, but
they do not get quite so far away from
home as that. They are very shy, too, and
the least strange thing frightens them.
'The woodland caribou,' says our naturalist, 'seems a wild, restless animal, even during the winter ranging through wide districts of country, and often changing his
home, and very suspicious and wary. An
alarm from which the moose would flee only
a few miles will send the caribou a whole
day at a rapid pace which takes him quite
out of the country and defies the pursuit
of the hunter.'"

"Don't they ever catch it, then?" asked
Malcolm. "I thought that the people in
Labrador ate it!"

"Yes, but it is said that only these people and other Indians can successfully pursue it, and it is very common for sportsmen
who go hunting in far regions to take an
Indian with them. 'It is in the damp and
fresh-fallen snow that the caribou is most

successfully followed. Then it is that the
foot clad in the moccasin made from the
skin of the hock of the moose returns no
sound to the hunter's step, and he is enabled to glide through the dark forest or
the bleak barren as noiselessly as a cat
upon a carpet. In districts where the caribou is not hunted except by the Indians,
as in the interior of Newfoundland and
Labrador, caribou are less suspicious and
less difficult to approach. There they have
their regular trails and run-ways, which they
pursue in their ordinary migrations, always
crossing the streams at favorite fords. In
these migrations the deer march in small
bands, in single file, generally several feet
apart, in well-beaten paths. Their march is
leisurely made and rather slow. They frequently pick the lichens as they pass, unless
they observe something to excite their sus-
picions. This is the time for the natives to
make their harvest of meat. The greatest opportunity is at the ford of a broad stream.'"

"Poor things!" said Clara. "It seems a
shame to kill them."

"You would not think so, perhaps, if

you were fond of venison and could not
get much else to eat. Sometimes, in winter, the caribou will cross on the ice the
water that separates Newfoundland from
Labrador; and when it does this the Indians can more easily waylay and destroy
it. They pursue the animals with bow and
arrow, and succeed in slaying a great many.
Once three Indians who had been watching
several hours for caribou encountered more
game than they expected. Hearing the
clatter of hoofs over the rocks, they looked in a direction from which they least expected caribou to come, and there were two
caribou pursued by a small band of wolves
and coming to the very place where they
were lying. 'They were not more than
three hundred yards away and were coming
with tremendous bounds, and were fast in-
creasing the distance between themselves
and the wolves, who had evidently surprised them only a short time before.
Neither Michel nor his companions had fire-
arms, but each was provided with his bow
and arrows. The deer came on; the Indians lay in the snow, ready to shoot. The

unsuspecting animals darted past the hunters like the wind, but each received an arrow, and one dropped. Instantly taking a fresh arrow, the Indians waited for the wolves. With a long and steady gallop these ravenous creatures followed their prey; but when they came within ten yards of the Indians, the latter suddenly rose. Each discharged an arrow at the amazed brutes, and succeeded in transfixing one with a second arrow before it could get out of reach. Then, leaving the wolves, they hastened after the caribou. There, quite close to that steep rock,' continued the interpreter who was telling the story, 'the caribou which Michel had shot was dead; he had been shot in the eye, and could not go far. Michel stopped to guard his caribou, as the wolves were about. One of his cousins went after the deer he had hit; the other went back after the wolves which ' had been wounded. The wolf-cousin had not gone far back when he heard a loud yelling and howling. He knew what the wolves were at: they had turned upon their wounded companion and were quarreling

over the meal. The Indian ran on, and
came quite close to the wolves, who made
so much noise and were so greedily devouring the first he had shot that he approached
quite close to them and shot another, killing
it at once.   The caribou-cousin had to go a
long distance before he got his deer.'"

"What horrible creatures wolves are!"
exclaimed Malcolm. "I'd like to kill five
hundred thousand of 'em."

"What an ambitious desire!" laughed
his governess.   "But wait until we come
to wolves; you may find them very interesting."

"But they chased the pretty deer," re-
monstrated Edith, "and wanted to kill 'em."

"And I'm very much afraid, little girlie,"
replied the young lady, "that the pretty
deer would chase you, and kill you too, if
they got a chance."

"Would they eat her?" asked Clara,
with great interest.

"I think not," said Miss Harson, "as they
are not carnivorous—that is, flesh-eating.
But it would make no difference to Edie
after she had been killed whether they did

or not, and it is best to keep out of their way."

"Yes," said Edith, very earnestly; "I'm going to run as soon as I see one."

"Climb a fence, Baby," suggested her brother, "or get up a tree; then you will be safe."

This was rather mean of Malcolm, as he knew very well that poor "Baby" could do neither one nor the other, and there was a very suspicious approach to crying which Miss Harson contrived to nip in the bud.

"*I* can't climb a tree," said the young lady, laughing, "and I am not at all sure about going over a fence; but, as I never expect to have to run away from caribou or other wild animals, this does not trouble me at all.—Now let us see what else travelers say about our American reindeer: 'The caribou travel in herds varying from eight or ten to two or three hundred, and their daily excursions are generally toward the quarter whence the wind blows. The Indians kill them with the bow and arrow or with the gun, take them in snares or spear them in crossing rivers or lakes.

The Esquimaux also take them in traps ingeniously formed of ice or snow. Of all the deer of North America, they are the most easy of approach and are slaughtered in the greatest numbers. A single family of Indians will sometimes destroy two or three hundred in a few weeks, and in many cases they are killed for the sake of their tongues alone.'"

"That seems dreadfully cruel," said Clara.

"It does indeed, Clara," replied her governess, "and this is often the case where wild animals are plentiful. 'The Esquimaux trap these deer, using the reindeer-moss for bait. The trap is constructed of frozen snow or ice enclosing a room of sufficient dimensions to hold several deer, and over this is laid a thin slab of ice supported on wooden axles forward of the centre of gravity. The top of this is accessible only by a way prepared for the purpose, and beyond is laid the tempting moss. In reaching it the deer passes over the treacherous slab of ice, which is tilted by the weight of the animal, and he is

precipitated into the room below, when
the top, relieved of the weight, resumes its
horizontal position and is ready set for
another victim. Great numbers are captured by the Indians by driving them into
pens or enclosures made of bushes and
placed in the course of some well-beaten
path where a narrow gateway is left, from
either side of which is placed a diverging
line of bushes or piles of stone, perhaps
one hundred feet apart. These may extend
a mile or two and at their extremities be
far apart. A watch is kept from some
high point of observation; and when a
herd of deer is observed approaching, the
whole family, men, women and children,
quietly skulk around them and drive them
within the lines of objects which in their
stupidity and on account of their defective
eyesight they regard as impassable barriers,
and so rush straight forward upon the path
into the enclosure, in which is a labyrinth
of ways made by rows of bushes, where
the deer become fairly dazed and are
slaughtered with spears, and even with
clubs, the women and children in the mean

time guarding the outside of the enclosure
to prevent the escape of any. The number slaughtered in this way is very great,
and furnishes the natives with provision
in great abundance.' This," added Miss
Harson, "although it sounds cruel, is not
so, because it is done to obtain necessary
food; and we shall see that wild animals
which can be used for food are always
most abundant where little else can be
found to eat."

"Can't deer see well," asked Malcolm,
"when they have such large, bright eyes?"

"No," was the reply; "their eyes are
beautiful to look at, but they seem to be
of very little use in keeping them out of
danger. They see things quickly, but they
do not appear to know what they are.
Their sense of smell is keen, but even this
may be overcome by cunning; and there
are a great many stories of the deceptions
used by various tribes of Indians to snare
and kill this valuable animal."

"Please tell us some more, Miss Harson," pleaded Clara.

"Please do," chimed in Edith, forgetting

her horror of "the cruel Indians who killed the pretty deer."

"Don't some of 'em dress up in the horns and skin?" asked Malcolm.

"Yes; that is practiced among the Dogrib Indians. 'The hunters go in pairs, the foremost man carrying in one hand the horns and part of the skin of the head of a deer, and in the other a small bunch of twigs, against which he from time to time rubs the horns, imitating the gestures peculiar to the animal. His comrade follows, treading exactly in his footsteps and holding the guns of both in a horizontal position, so that the muzzles project under the arms of him who carries the head. Both hunters have a fillet of white skin around their foreheads, and the foremost has a strip of the same around his waist. They approach the herd by degrees, raising their legs very slowly, but setting them down somewhat suddenly, after the manner of a deer, and always taking care to lift right or left feet simultaneously. If any of the herd leaves off feeding to gaze upon this extraordinary phenomenon, it instantly

stops, and the head begins to play its part by
licking its shoulders and performing other
necessary movements.   In this way the hunters attain the very centre of the herd with
out exciting suspicion, and have leisure
to single out the fattest.   The hindmost
man then pushes forward his comrade's
gun; the head is dropped, ànd they both
fire nearly at the same instant.   The deer
scamper off; the hunters trot after them.
In a short time the poor animals halt to as-
certain the cause of their terror; their foes
stop at the same moment, and, having loaded as they ran, greet the gazers with a second fatal discharge.   The consternation
of the deer increases; they run to and fro
in the utmost confusion, and sometimes a
great part of the herd is destroyed within
the space of a few hundred yards.'

"I think, Malcolm," said his governess,
smiling, " that you will know now exactly
how to act when you go North to hunt the
caribou. With a few words more about the
Esquimaux way of capturing this animal,
I am sure you will need no further instruction: 'When feeding on the level ground,

an Esquimaux makes no attempt to approach him, but, should a few rocks be near, the wary hunter makes sure of his prey. Behind one of these he cautiously creeps, and, having laid himself very close with his bow and arrow before him, imitates the bellow of the deer when calling each other. Sometimes, for more complete deception, the hunter wears his deerskin coat and hood so drawn over his head as to resemble in a great measure the unsuspecting animals he is enticing. The bellow is very attractive; yet if a man has great patience, he may do without it and be equally certain that his prey will finally come to examine him, the reindeer being an inquisitive animal, and at the same time so silly that if he sees any suspicious object which is not actually chasing him he will gradually, and after many caperings and after forming repeated circles, approach nearer and nearer to it. The Esquimaux rarely shoots till the creature is quite close. The great curiosity of this deer leads it to destruction, and it has not the acute sense of smell which is possessed by the other deer. Add to these infirmities its

stupidity and the fact that it is easily dis-
tracted, so that it is incapable of escape
even in the open plain, and we have the
picture of an animal which is very useful to
the natives, who have to depend on the rud-
est and most imperfect weapons to procure
subsistence.' "

The little Kyles enjoyed these accounts
so much that Miss Harson said she would
have to send them off among the Esquimaux. Edith was almost frightened and
Clara made a wry face, but Malcolm declared it was just what he wanted—for a
while.

CHAPTER X.

THE children were laughing one afternoon over the picture of a very absurd and not at all amiable-looking animal which they called a "big reindeer." Perched high up, as though he had been on stilts, this queer creature looked as if he had humped up his shoulders to reach his horns, and these were of a very funny description, being more like immense leaves than anything else.

"This outside one is like a hand in a mitten with a very long thumb," cried Malcolm, excitedly. "Look at it sticking out there!"

"Not a bad description," said Miss Harson, who came into the room while he was speaking, "and these queer antlers are called 'palmated,' because they do bear some resemblance to a hand or a palm. They are uneven, you see, and those on the other side are shorter, looking like two hands with only three fingers apiece. The reindeer's antlers are slightly palmated at the ends, but only the moose—for that is the name of this beautiful being—has antlers like boards with jagged edges."

" I think," said Clara, carefully examining the long, ungainly head, " that he looks like a pig."

" He's got a regular horse's head," said Malcolm. " Look at his nose!—Which of us is right, Miss Harson?"

" Both," was the reply. " I think he suggests both a pig and a horse.—And what does Edie say?"

" He looks queer," replied the youngest naturalist. " I'm glad he isn't alive."

All laughed at Edith's remark, and agreed that it would not be altogether pleasant to have such a great lumbering creature prowling about at Elmridge.

" How low his antlers are!" said Malcolm. " They look so spread out."

" Yes," replied his governess, " and they

also look very strong, making formidable weapons of warfare. 'The moose,' says a naturalist, 'like all the others of his genus, joins battle with a great rush which must often try the strength of the antlers to the utmost; yet we have no account of the antlers being broken short off, but it frequently happens that the tines or snags are dislocated. But for the great elasticity possessed by antlers over all other bones, owing to the larger proportion of animal matter which they contain, a single battle would serve to destroy them.'"

"What are 'tines or snags,' Miss Harson?" asked Clara.

"Why, you know what the tines of a fork are—the divisions. You see these even in the antlers of the moose, on the edge. I should like to have you remember as well as you can that the main stem of all antlers is called the 'beam;' the larger branches from the beam are called 'tines,' and the branches from these, and small branches from the beam, are called 'snags.' The flattened portions of either the beam or the tines are called 'palms.' There is

14

more to be learned about antlers, which can be left for some future time."

"I don't believe," said Malcolm, ruefully, "that I shall ever remember anything but 'tines,' because I can think of forks for that."

"If that will help you so much," said Miss Harson, "you can think of the Mississippi River for snags, and of a verse in the Bible for beam, and of your hands for palms. But it always seems to me easier to think of the thing itself in the beginning."

"That's because you can think of anything you want to, ma'am," replied her oldest pupil, "but I'm like that boy who had such 'a good forgettery.'"

"Nonsense!" said the young lady, laughing. "A bad memory is often only want of interest. Really care for things which you ought to remember, and you will find the remembering quite easy. Now let us see what is said of the appearance of our handsome new friend: 'In form the moose is an ungainly animal—a short body, a very short tail and neck, a prodigiously long, ugly head, with a projecting nose or

upper lip, which gives the animal a revolt-
ing look.   He has enormous ears, short,
spreading, palmated antlers and very long
legs, to which he is indebted for his great
height.'   He is about the size of an ordinary horse, but not nearly so graceful;

MOOSE.

and in addition to all these charms he has
what is called 'a pendulous appendage'
—which means something hanging—under
his throat.   It is covered with long, coarse
black hairs, and hunters speak of it as 'the
bell.'"

This description just matched the picture,

which was now examined with fresh interest.

"What color is a moose, Miss Harson?" asked Clara.

"Very often of a jet black, if it is an old bull-moose; when it is younger, its color is a tawny brown on the upper side, but much lighter underneath. The cow-moose is of a light sandy color above and almost white beneath. The calves are of the same hue, rather indistinctly spotted. The color of this animal varies somewhat with the season, and so does its coat. In summer this covering is of soft, fine, firm hair, while the winter coat, which is at first short, fine and glossy, as it gets later grows coarse and open. For midwinter the moose is provided with a thick under-coat of fur."

"Does it come off and on?" said Edith, who was thinking of her doll's clothes.

"Not in the way you mean, dear," was the laughing reply, "as it would be rather difficult for so awkward an animal to dress and undress itself. The fur comes out as hair does when summer approaches, and grows in again as soon as it is needed."

"These ears," said Malcolm, "are the queerest-looking things! See what tremendous ones the cow-moose has."

"Yes," replied his governess, "they are nearly a foot long, and very broad besides. The eyes are small for an animal of the deer family, and they are said to be capable of a most malignant expression. See, too, how square the muzzle is and what a deep cleft it has. The upper lip stands out several inches beyond the lower, and is called 'prehensile,' which means movable and able to hold fast. 'With this organ the moose is able to hold on to the boughs and twigs of tall saplings '—which is the kind of food he particularly likes—'and to convey them within the grasp of his powerful teeth.'"

"Doesn't he eat moss, like the reindeer?" asked Clara.

"Not when he can get young trees and shrubs, as his long legs and short, thick neck make it hard work for him to feed on anything so low. He is the largest of all the deer species, and belongs to the family of elk or wapiti; but the animals especially called by this name are smaller and in some respects different."

"Are there any mooses near us, Miss Harson?" asked Edith.

"'Moose,' dear, not 'mooses.' No, there are none at all near us, for they like cold places in which to live. They used to be found in the Adirondacks—which, as I have told you, are in the northern part of New York State—and they were frequently found in Maine. Now one is occasionally seen in the wildest portions of the latter State, but they have been so persistently hunted and destroyed that they are a rare sight. I remember once," added the young lady, "when I was spending the early autumn in Maine, that I gathered some beautiful berries to put in a bouquet, and my friends there told me these were 'moose-berries.'"

"Oh!" was the eager exclamation. "What did they look like, Miss Harson?"

"Almost exactly like wax. They were small, long-shaped and in clusters. Not being fully ripe, they were just turning an exquisite cherry-color, and the part not

tinted was of a greenish white. They
grew on tall bushes, and I suppose that
the moose are in the habit of eating similar berries, with the twigs on which they
are found. They seemed to me the prettiest things of the kind I had ever seen.
The moose is found in New Brunswick
and Nova Scotia and the ice-bound regions
around; he is also very much at home in the
northern parts of Europe. These animals
have a curious habit of making settlements
for themselves in winter; these settlements
are called 'yards.' When the snow becomes deep in the forests which they inhabit, they will gather in small bands and
work industriously at these winter residences.
Some of them are much more complete than
others, but all are made by tramping the
snow down to a hard floor through the yard,
leaving it surrounded by a high wall of un-
trodden snow. 'The places selected for
these yards are dense thickets affording the
greatest abundance of shrubbery yielding
their favorite food. This they utterly destroy within their yard by consuming the
twigs and stripping off the bark. Even from

the large trees which they cannot bend
down in order to reach their tops they strip
the bark as far as they can reach. If they
do not relish this coarse, dry bark of the
large trees, they consume it all to satisfy
their hunger. When all the food within the
yard—which sometimes becomes considerably extended to reach the shubbery—is
gone, they break their way to another lo-
cation, where a fresh supply may be found,
and form a new yard.' Sometimes the yard
itself is small, but paths are made and well
packed down ; between the paths reaching
to the trees and shrubbery in the neighborhood the deep snow is undisturbed. The
moose feeds with great satisfaction upon
evergreens, and he is the only animal that
is known to do this."

"Is he good to eat?" was Malcolm's next
inquiry.

"The moose-hunters think so," was the
reply, "but the venison is rather coarse.
The upper lip is considered a great delicacy
by the natives, and the tongue is much es-
teemed by them. The skin is used for tent-
covers and shoe-leather, and there is so

much of it altogether that it is well worth hunting."

"I should think people would be afraid of those dreadful horns," said Clara. "I shouldn't like to go near it."

"Moose-hunting is dangerous work, and 'instances have been known in which the hunter has met his death in these encounters, his ribs being fractured by the powerful blows of the fore feet, and his whole body gashed and torn by the antlers of the infuriated animal.' With those immense ears the animal can hear at a great distance, and can as easily detect danger by the power of scent in the huge nose. 'The slightest crackling of a dried stick beneath the hunter's foot, the rustling of the underwood against his person, are conveyed to a great distance in the forest, and apprise the wary moose of danger.' Times of deep snow are preferred by the Indians for pursuing the animals, as then they are quite easily caught in the 'yards' and killed by wholesale. A single one can successfully be followed over the snow by an Indian on his snow-shoes, as the Indian can get on very rapidly, while the poor heavy moose struggles through the snow, at every step sinking in up to his thighs. He flounders out again and keeps on as long as he can, but it is terrible work even with his great strength, and before long he is quite exhausted. Later in the season, when the top of the snow is softened by the sun during the day and frozen hard during the night, a crust is formed which will bear a dog or a man, but not so heavy a body as that of a moose. A chase when the snow is in this condition is sure to end in favor of the hunter. The small, sharp foot of the animal cuts directly through this crust, and he sinks at every step, while in rising from it the sharp edges of the icy crust cut and bruise his legs, so that he cannot get on very fast and is soon overtaken and killed. This kind of hunting is called 'crusting.'"

"That seems mean," said Malcolm; "the poor moose doesn't have half a chance."

"I quite agree with you," replied his governess, "and catching the animal with a noose made from the hide of one of his relations is not much of an improvement.

HUNTING THE MOOSE.

'This was placed across a convenient limb which was suspended directly over the path in the forest, large enough for the head and antlers of the largest moose to pass through, but sufficiently high from the ground to answer the purpose. To the other end a heavy weight—usually a log of wood—was attached. This was held suspended high above the ground by a trip properly arranged, which was to spring by the least strain from the loop of the thong. Through this the moose would unsuspectingly pass till his breast or fore legs should touch the lower line of the noose, when the trip would be drawn tightly around the neck of the animal. A few minutes struggling and rearing must always end in his death. In this way the Indians captured many moose, elk, and other animals, before they obtained fire-arms.' "

"Do they catch 'em in any other way?" asked Clara.

"Oh yes; the Indians have various plans and stratagems for securing such animals as they need. It was not easy to kill so large an animal as the moose with bow and arrow unless they were very close to him; and with such ears to warn him of the slighest sound, and such a nose to inform him of the faintest unfamiliar smell, it seemed impossible to approach him. 'In summer-time he was more frequently captured in the water. At that season he affects marshy grounds where lakes and lakelets abound, and into these he plunges to escape the torments of the flies and mosquitoes, deeply immersing himself much of the time, generally with only his nose above water. In this position he could successfully be attacked by the Indians in their canoes at sufficiently close quarters to make their arrows effective, or they could even disable him with blows before he could escape. This was often dangerous sport, or business—whichever you please to call it—for a single blow from the antlers or the foot of a moose was sufficient to sink a canoe, when the hunter would be fortunate if he escaped with his life. This mode of pursuit was, however, generally successful, and much meat was obtained in that way by the natives.'"

"Miss Harson," asked little Edith, who had been deeply considering the matter, "do Indians and Laplands, and all such funny people, just eat meat for dinner, and nothing else?" It seemed to her that they were always hunting and killing things, but nothing was said about their raising vegetables or making nice bread and cake.

"Very often, dear," was the reply, "they have nothing else to eat. They live in cold, barren countries, you must remember, where it would be impossible to make things grow if they tried ever so hard, and *because* they need them God has made plentiful the very animals they need. It is God, too, who gives them the knowledge to catch these animals, and what would be very cruel for us to do is not cruel for them. Do you remember our talking on this very subject?—And these very moose whom you pity so much, Malcolm, often kill one another without any help from the Indians."

"What do they do that for?" asked the children, in great surprise.

"Because they are so quarrelsome; and

when two of these great creatures rush at each other in a fury, they always fight with their antlers, each trying to inflict a terrible wound upon the other. A naturalist says that when wandering through the woods he has several times found the skeletons of two moose whose antlers had become so firmly interlaced in their encounter that, unable to extricate themselves, the animals had miserably perished face to face. Other deer have been found entangled in the same manner. Even a young moose is very ferocious, and Audubon, the great American naturalist, says of one that had been captured, 'The moose was so exhausted and fretted that it offered no opposition to us as we led it to the camp, but in the middle of the night we were awakened by a great noise in the hovel, and found that, as it had in some measure recovered from its terror and state of exhaustion, it began to think of getting home, and was much enraged at finding itself so securely imprisoned. We were unable to do anything with it; for if we merely approached our hands to the openings of the hut, it would spring at us with the greatest fury, roaring and erecting its mane in a manner that convinced us of the futility of all attempts to save it alive. We threw to it the skin of a deer, which it tore to pieces in a moment. This individual was a year-ling and about six feet high.'"

"Just think," exclaimed Clara, "of being six feet high and only a year old!"

"This would be strange indeed for a human being," replied her governess, "but it is not strange for a moose. One peculi-arity in which this animal indulges is quite amusing: when it hears a sudden noise and starts on a run, it will sometimes fall down suddenly, as if it had been taken with a fit. This does not last long, how-ever, and presently it is up and off again, shambling along in its clumsy way, but getting swiftly over the ground."

"Miss Harson," said Malcolm, "do the people where the moose live ever catch 'em and make 'em draw things, like the reindeer? I s'pose they'd run away and upset everything, wouldn't they?"

"It is not easy to tame them and make

them useful, but it seems that this has been done. They are only about half tame, though, at best; and a writer speaks of seeing one, when a boy, that was kept in a barn and would attack any one within reach that seemed to be afraid of it. The moose has sometimes been broken to the harness and made to draw heavy loads; and were it not for the 'wicked disposition' which never seems to be subdued, it would prove a useful beast of burden in regions where there are heavy snows. As it is, this formidable animal is much more useful dead than it is when alive."

15

CHAPTER XI.

THE WAPITI, OR ELK.

"ANOTHER very large deer," said Miss Harson, "is the wapiti, or elk, which is next in size to the moose, it being over five feet high and often weighing a thousand pounds."

"Why, Miss Harson!" exclaimed Clara, in a tone of awe; "that's more than all of us put together."

"I hope so, dear," was the laughing reply, "with papa and John and Thomas and Kitty thrown in. But animals are always heavier in proportion to their size than human beings, and the elk is by no means so large as all of us put together. It is a much handsomer animal than the moose, although, as a writer says, 'one is not struck with its beauty when it is listlessly standing in some retired shade quietly ru-

minating ; but when awakened by excitement, it seems to change its form : anima-
tion and expression pervade every feat-
ure of the animal, and we are at once
charmed by a beauty and a symmetry which
before were entirely wanting. The spring
coat in which this animal appears is very
fine and glossy, of a deep cream-color, or
écru, with chestnut-brown on the legs, neck
and head. It fairly glistens in the sunshine,
but its appearance is often spoiled by the unwillingness of the old coat to give place to
the new one. During the winter this old one
gets so matted together that it is like a piece
of thick felt, and, instead of dropping off, as it
should do, it is torn away in great patches
by the forest-twigs, and basketsful of it
could be gathered in a small space. 'The
contrast between the new spring dress
which may perhaps appear on a part of
the animal, and the other portions, which are
covered with the shaggy and tattered winter
dress hanging about in torn patches, some
dangling a foot or two from the body, is indeed quite remarkable. The one seems em-
blematic of poverty and destitution, while

the other looks like thrift and comfort. One appears like the fag-end of a hard winter, while the other suggests the freshness and the gayety of spring.' "

"What very funny things animals do in the woods!" said Malcolm. "I think they might sometimes give a fellow a chance to see 'em when they are at some of their queer antics."

"They can scarcely be expected to give 'a fellow' who is hundreds of miles away such a chance," replied his governess, "but the 'fellow' who wrote what I have just read you put himself in the way of seeing them by going where they were. Naturalists discover a great many curious things of which the rest of the world know nothing.—In winter the color of the elk is a dirty white, with black underneath, and he is royally crowned with magnificent branching antlers—'the longest, most graceful and symmetrical antlers of all known deer.' They are also the most formidable as weapons and shields, and sometimes measure as much as five feet in length."

"Oh!" exclaimed the children, while

Clara asked if that was not even worse than the moose.

"They are certainly much longer," continued Miss Harson, "and the 'charge,' as it is called, of an elk when he makes a rush with his formidable antlers is a rather serious matter. It seems that, in moving, a herd of these animals 'will start at first quite leisurely; presently one or two will strike a trot, when all will do so, except the young ones, which break into a run. The pace is increased by all till they reach a bluff or a ravine, when all break into a furious run and come thundering down the cliff like an avalanche. When you see forty or fifty elk, more than one-fourth of them having huge antlers, come rushing down toward you, you feel glad there is a good fence in front of you.' Some of them go very fast, although their usual gait is a trot; and it is said that an elk will trot across an ordinary prairie at the rate of a mile in a little over three minutes."

This was quite exciting to Malcolm, who seemed ready to go at once in quest of a prairie and an elk.

"What does he eat, Miss Harson?" asked Edith, remembering the moose. "Are there any pretty elk-berries?"

"No, dear," was the smiling reply; "I have never heard of any; but here is something that will tell us all about it: 'The wapiti deer selects his food from the trees and the shrubs, the grasses and the weeds, though he is not so fond of the latter as some of the others. Like several of the other species, he prefers the bitter and the astringent, like the hickory and the oak, to the hazel and the maple. He may often be seen standing erect on his hind feet, stretching his neck to the utmost to get a bunch of leaves nearly beyond his reach. In the winter he frequently pulls down the twigs bearing the dry oak-leaves, and eats them with apparent relish, though he is rarely seen to pick up those which have fallen after maturity. If deprived of arboreous'—'tree' or 'shrub'—'food, he will keep healthy and fat on grass alone. In winter he will scrape away deep snow with his feet to obtain the grass beneath it, and by some unexplained means seems always to select the best places.' These animals are great eaters, and are not at all particular about what they eat, according to those who have kept them in parks. They will live in winter on cornstalks, and will even eat damaged hay which horses and cows would refuse. They will not take the trouble to find food for themselves so long as anything to eat is put in their way. They like all kinds of grain, and an enormous ear of corn will be crunched up at once, cob and all."

"Do they get real nice and tame like cows?" asked Clara.

"Not exactly like cows, dear; for when you hear of some of their ways, you will scarcely think them so gentle as that. When they are entirely dependent on being fed, they will come at the call of the person who feeds them; but when fresh grass can be had, they are not so obedient. They are not pleasant-tempered animals, and a mother-elk will often treat a helpless little one not her own with great cruelty, sometimes killing it outright just because it happens to get in her way. But they are very careful of their own fawns, and

will seldom allow them even to be looked
at.   A gentleman who had a herd of these
animals says: 'I was once driving through
the park, when we observed an old doe
whose anxious look excited suspicion.   We
hitched the horses and commenced a search
for a fawn; at last we saw it curled up in
the leaves, perhaps two hundred feet from
the dam, who faced us all the while.   When
she saw we had discovered it and were
going toward it, she uttered a succession
of threatening squeals which sounded to
us anything but musical, at the same time
walking slowly toward us with a gleam of
the eye and an air not to be mistaken.
We did not count the spots on that fawn
that day, but retreated in as good order as
possible with our faces to the foe.   My
friend, who was not used to the animal,
remarked while I was admonishing him to
show no signs of fear, but to retire as if it
was quite voluntary, "I would give a big
check to be in that buggy now."   Had we
run from her, we might not have won the
race without trouble.'   Another doe belonging to the same gentleman was much

more tame, as it had often been fed from his hand; and one day, when he found her carefully licking a young fawn, she seemed quite willing to have him pet it and lift it on its feet. She did not appear to think of such a thing as his hurting her baby, when he had always been so good to her, but stood looking on with great satisfaction. Yet the writer was obliged to add: 'But the amiable one was not always amiable, and not always to be trusted. I once came across her when rambling through the park with my little daughter. I left her feeding the elk and walked away, perhaps to pick some wild flower, and turned round just as the brute struck at the child. Fortunately, she was not quite in reach. I spoke to her in no very mild terms, and the blow was not repeated. There was manifested a disposition to strike the child simply because she knew it was unable to protect itself.'"

"Then," said Edith, with great indignation, "elks are hateful, wicked things, and I don't want papa to have any at Elmridge."

"I did not know that papa had any such idea," replied her governess, laughing, "and I certainly should not enjoy having them about these grounds. But we will not worry about it, Edie, for I feel quite sure that this trouble will be spared us."

"It's so horrid," said Clara, "of the old does—or whatever they are—to hurt the poor little fawns! They ought to be killed themselves."

"I'd like to go after one of 'em," exclaimed Malcolm, who was also indignant, "with my bow and arrows. I don't see why any one wants to keep such pets as those."

"People certainly do have queer fancies in this respect," said Miss Harson, "and all vicious animals should certainly be killed. But I believe that the naturalist who kept these elk particularly wished to study their habits. Some others of the herd were so ferocious that they could not with safety be approached at any time."

"Are the little fawns pretty?" asked Clara.

"Their heads and faces are," was the reply, "and they have very neat little hoofs; but

their legs are as long in proportion to the
body as those of a young calf, and this
gives them an awkward look. They are
prettier when they are lying down, and a
little curled-up fawn on a heap of leaves
is certainly very 'cunning.' But—would
you believe it ?—these little creatures, when
only a few hours old, will 'make believe dead'
if any one goes near them. The funniest
part of it is that they do not seem to know
enough to shut their eyes, but they are as
still as possible. The same naturalist says:
'They lie without a motion; and if you
pick them up, they are as limp as a wet
rag, the head and limbs hanging down
without the least muscular action, the bright
eye fairly sparkling all the time. The first
I saw really deceived me, for I thought it
had met with some accident by which it
was completely paralyzed, and I returned
the next day expecting to find it dead. It
was gone, and soon afterward I found it
following its dam in as sprightly a manner
as possible. Last spring I found one,
picked it up, carried it some distance and
laid it down, and watched for a while.

But not the least sign of life would it manifest, save only in the bright eye.'"

The children were wonderfully amused with this account, and Edith felt "so sorry that the poor little fawn *would* forget to shut its eyes." Animals were certainly getting funnier and funnier.

"The same thing is done by Papa Elk and Mamma Elk," continued the young lady, "only in a different way; and you will see that there is great danger in hunting this animal. Here is an account of a hunter who fired at one that was standing on a ledge of rock overhanging a deep pool and about thirty feet above the water. The deer dropped, and the hunter hastened to secure his prey. He grasped it by the horns, and was just about to make sure of it with his knife, when the animal suddenly sprang up and began a ferocious attack upon him, evidently intending to drive him into the water. The elk had made such a powerful spring that both tumbled from the rock together and into the pool below. The fallen hunter now cared for nothing but to gain the bank again, and the animal, too, had evidently not calculated upon a ducking for himself. As the man reached the shore his prey was just disappearing in the distance, but some time afterward he shot a deer which from the scar of a wound in its neck he recognized as his former assailant."

"He must have been glad," said Malcolm, "to finish him at last, after having been knocked into the water by him."

"I have no doubt he was glad," replied Miss Harson, "for such is human nature ; but the elk was, after all, only trying to preserve his own life. He did not make the attack. This animal has also another kind of cunning, which is practiced for its preservation, and that is to hide itself among surroundings that are very nearly its own color. It often escapes in this way, for it is said that only an experienced hunter can see where it is lying, even when the exact spot is pointed out. Elk are most easily secured by watching for them near the salt marshes, or 'licks,' where they go in great numbers to satisfy their craving for salt kinds of food. The hunters conceal themselves near the salt-licks, and by watching their chances kill great numbers of the deer as they pass to and fro."

"What do they do with them after they are killed?" asked Clara.

"The skin is valuable, as it makes soft leather, and the antlers are ornamental; but the flesh is coarse, and the tongue is the only part that can be called good eating."

CHAPTER XII.

MISS HARSON had taken a large engraving from the portfolio in the parlor, and was showing it to her little charges. The picture was called "The Stag at Bay," and the children were full of pity for the beautiful animal that stood partly in the water, panting and trembling and foaming at the mouth, beset by two fierce-looking dogs, one of which he had felled to the ground and wounded, if not killed, with hoofs and antlers, while the other showed his cruel fangs and did not seem inclined to approach any closer.

"Miss Harson," asked Malcolm, "what does 'at bay' mean?"

"It means," replied his governess, "hemmed in so that it is impossible to escape, and yet resolved to fight to the very end.

239

This noble animal will sell his life as dearly
as possible, and perhaps will kill the other
dog too before the hunters, who are not far
off, come up and despatch him."

"How wicked that is!" exclaimed Clara,
indignantly.

"It certainly *is* wicked," continued the
young lady, "to put an end to any creature's life merely for the pleasure of the
sport. But this is not always the case, as
the red deer, or stag, of Europe and our
own American deer are good eating. The
animal in the picture is one of the former
kind, and the scene is among the wild 'lochs'
of Scotland. These lochs are sheets of
water surrounded by steep hills, and such
a region is a favorite resort of deer. It is,
as you see, a very pretty animal with a
small nose and mouth, and with ornamental branching horns not unlike those of the
wapiti. These horns can do a terrible
amount of mischief when the animal is
brought to bay, as during a chase one deer
has frequently killed several dogs, and has
threatened the hunters besides. These
animals are of a reddish-gray color and

only about half the size of the elk, although some of them are as large as a small elk. The size varies almost as much as in human beings. The common deer of this country are still smaller, and are really another species; but they are alike in so many things that it is not necessary in our simple talks to take them up separately. Both are very shy, yet both can be tamed; and if captured when very young, they become contented and affectionate."

"Do they fight like the moose and the elk?" asked Malcolm, who never was willing to miss a chance for hearing accounts of battles.

"They certainly fight," was the reply, "for all deer do that, but their battles are not so fierce. I do not mind treating you to a story of one, Malcolm, as it is not a bloodthirsty one, and the account is just where I can lay my finger on it. Two large bucks, it seems, were kept in adjoining parks, and 'after their antlers had become hard they occasionally saw each other on opposite sides of the fence, when they would make faces at each other, with vari-

16

ous threatening demonstrations, showing
that both were ready for the fray.'"

That deer should actually "make faces at
each other" was perfectly delightful, only
it did seem too bad to the children that
they were not there to see.

"One day the owner of the two parks
ordered the passage between them to be
opened, and the result was a terrific fight.
'The battle was joined by a rush together
like rams, their faces bowed down nearly
to a level with the ground, when the clash
of horns could have been heard at a great
distance; yet they did not again fall back
to repeat the shock, as is usual with rams,
but the battle was continued by pushing,
guarding and attempting to break each other's guard, and by goading whenever a
chance could be gotten, which was very rare.
It was a trial of strength and endurance
assisted by skill in fencing and by activity.
The contest lasted for two hours without
the animals once being separated; during it
they fought over perhaps half an acre of
ground. Almost from the beginning both
fought with their mouths open. So evenly

matched were they that both were nearly exhausted, when one suddenly turned tail and fled: his adversary pursued him but a little way. I could not find a scratch upon either sufficient to scrape off the hair, and the only punishment suffered was fatigue and a consciousness of defeat by the vanquished.'"

"What do they want to fight for, Miss Harson?" asked Clara.

"A pleasant little habit they have 'to decide which is the better deer;' and if they happen to be separated for some time, a fresh battle is necessary before one will acknowledge the superiority of the other. It does not matter so much when they don't get hurt, but it must be quite absurd to see one of these encounters. Animals certainly are very funny and much more interesting than those who know little about them would suppose."

"Miss Harson," said Edith, who was looking at the stag in the picture, "will not this poor deer get drowned in the water?"

"No, Edie, for the deer can swim, and taking to the water is often his only chance

for life. I have been reading an account
of a splendid deer pursued by a pack of
wolves; it was seen by some hunters and
their Indian guide to dash from a thicket
into some shallow water covered with lily-
pads and rush through it, but more slowly
as the water deepened. 'When he reached
the edge of the lily-pads and the deep clear
water was right before him, he stopped short,
threw high his head, displaying to the best
advantage his great branching antlers, and
looked back and listened at the yelping
of his pursuers.   There stood the monarch of the forest in the border of the quiet
lake, where the deep solitude is rarely broken by invading man, not dreaming there
were enemies before him more dangerous
than those behind, of escape from which he
now felt assured.' "

   "Oh what a shame!   Didn't the poor
deer get off, after all?"

   " No; the noble animal dropped into the
deep water and swam directly toward his
human enemies, who were in a canoe.
Presently the deer discovered them, and,
turning again to the shore, 'swam like a

STAG CHASED BY WOLVES.

racehorse.' The canoe, guided by the Indian, flew through the water all around the
deer to prevent his escape. He was finally
shot and dragged to the hunting-camp on
shore. 'It was slow work,' says the hunter
who shot the splendid animal, 'towing the
deer through the lily-pads, which extended
out for fifty yards or more. Before we
landed the three Indians on shore rushed
into the water, seized and dragged the deer
to the bank. He must have been a great
warrior, for all the points on his antlers
were broken off. He was a big deer, and
was a beautiful sight as he lay there upon
the green grass.' Here is an interesting
picture of a beautiful stag chased by wolves.
He is a pitiable object to look upon. See
how every muscle is strained in flight! His
tongue hangs from his panting mouth, and
his knees are bent for one frantic spring, in
the vain hope of eluding his terrible pursuers. But it is of no use: their sharp
fangs will soon sink into the poor stag's
flesh. Indeed, one of them has already
seized him by one of his flanks."

"I suppose," said Malcolm, "that the

people who first came to this country had to kill deer and such things to eat, didn't they, Miss Harson?"

"Yes," replied his governess; "the first white settlers, and the Indians, or aborigines, before them, had largely to depend for food on what they could catch and kill in the forests and the streams around them. A deer in those days was a great prize, and both Indians and white men became quite expert in hunting this animal. One of their methods—still in use—was called the fire-hunt. This took place at night, when the deer is apt to be roving about 'in the farmers' grainfields, around salt-licks or along the margins of rivers.' Generally two persons go upon a hunt of this kind; one of them carries a torch of pitch-pine above his head, while the other, with the gun, walks in front or behind. They must proceed with great care and watchfulness and without making the least noise. 'The deer sees the light slowly approaching and is rather fascinated than alarmed by it, and so he faces and stares at it in wonderment, when his eyes act as mir-

rors and reflect back the light, and appear
to the hunters as two great stars—or, as
they sometimes express it, like two balls of
fire set in nothing but darkness ; but neither
of these expressions gives a correct idea
of the appearance of the light reflected by
the animal's eye. The radiation of the
star is not seen, and the light is white instead of being the red light of fire. Nothing else of the deer is seen.' The poor
animal is, of course, shot as soon as the
hunters are near enough to shoot him ;
and if no noise has been made, the deer
seems unable to move, owing to the fasci-
nation of the light. Another trick with
Indian hunters was to dress in the skin of
a deer, with head, antlers and all, and
then, closely imitating the motions of the
animal when feeding, to get among a herd
and select a fat prize."

"I shouldn't think there would be any
deer left," said Clara, "when people have
so many ways of killing' em."

"The herds have been very much thinned
out," was the reply, "and entirely banished
from places where they used to roam ; but

there are still a goodly number left. One does not often see them here as tame as they are in some English parks, where the beautiful fallow-deer wander about or lie down in groups under the shade of the trees. They are usually very gentle, and will come and eat food from a person's hand; but occasionally there is a violent one who will attack those whose appearance he does not like. An account is given of 'a gentleman, an amateur in landscape-drawing, who had ventured into a park heedless of danger, and was engaged in sketching, when a deer saw him and charged full upon him. Down went his pencils and his papers, and he was only too happy to escape the animal's fury with the loss of his drawings. The creature's assault had a beneficial effect, for it taught this draughtsman an art of which he had thought himself ignorant. Hearing the animal close behind him, he seized a branch that hung overhead and curled himself into the tree with an activity that could be expected only from one versed in the practice of gymnastics.' "

"Wasn't he surprised?" asked Edith, in such an innocent way that the others thought her question quite as funny as the story.

"He certainly was, dear," was the laughing answer, "and I do not believe that he ever again went into a park to draw without being quite sure that there was no danger of meeting a deer. There is another story of tame stags, where the animals themselves were most unexpectedly chased. An English nobleman had trained four stags to draw a chariot, and was very proud of his curious steeds; but once the whole equipage came near being destroyed. 'The nobleman was driving to Newmarket, when the cry of a pack of hounds burst upon his ear. Unfortunately, the hounds came across the road over which the four stags had just passed. The hounds immediately changed their course and set off at full speed after the stags, whose scent was too great a temptation to be resisted. The stags, on hearing the cry of the dogs, bounded off at their swiftest pace, in spite of the efforts of the driver and the mounted grooms who always accompanied the equipage. The pace grew more and more furious on the part of pursuers and pursued, and the driver began to fear for the safety of his vehicle and himself, when he bethought himself of an inn at Newmarket where he had been in the habit of stabling his horned steeds. To this inn he directed all his efforts, and fortunately succeeded in getting his vehicle within the gates. The stags were now overpowered by the united force of hostler and stable-boys, and the whole party—stags, vehicle and driver— were thrust into a barn and the door shut just as the hounds entered the innyard.'"

"Do you think, Miss Harson," asked Clara, in some anxiety, "that any dogs would be likely to run after Prance and Caper when we are driving them?"

"Why, no, little girlie. Dogs are not trained to chase goats, but it is the especial business of hounds to go in pursuit of deer; so you must not think of asking papa for a deer-carriage."

There was not much danger that either of the little sisters would want such a present as this, but Malcolm appeared to think that it would be a fine thing to drive a pair of stags in harness. Miss Harson said that he could not be allowed such a team unless he would agree to take four, and here for the present the matter dropped.

CHAPTER XIII.

DEAR LITTLE DEER.

"MISS HARSON," said Clara, "I haven't found anything about deer in the Bible, but I looked all over for 'em."

Malcolm had the same report to make, and their governess said that she was not at all surprised.

"You did not find them," continued the young lady, smiling, "for the best of reasons, because they are not there—at least, not under that name; but there are places where the hart and hind and roebuck are mentioned, and the hind is supposed to be the same as the fallow-deer of English parks. 'As far as can be ascertained,' says a writer on Bible subjects, 'at least two kinds of deer inhabited Palestine in the earlier days of the Jewish history, one belonging to the division which is known by its branched horns, and the other to that in which the horns are flat or palmated over the tips.' The hind is first mentioned in Genesis, forty-ninth chapter, twenty-first verse, where Jacob blesses his sons and says, 'Naphtali is a hind let loose.'"

"I wish I had known that," said Clara, "for then I could easily have found the places."

"And had I known that you were looking for them, dear," replied Miss Harson, "I could have helped you, but I see that you wished to surprise me. You shall find some of the places now, if you like—both of you—and first turn to the First Book of Kings, the twenty-second and twenty-third verses of the fourth chapter."

The little girl wondered as she began the twenty-second verse, but she read very reverently:

"'And Solomon's provision for one day was thirty measures of fine flour, and three-score measures of meal, ten fat oxen and twenty oxen out of the pastures, and an hundred sheep, beside harts, and roebucks, and fallow-deer, and fatted fowl.'"

"Miss Harson," asked Edith, very solemnly, "was King Solomon a giant?"

"Why, no, dear child. What ever put such an idea into your little head?"

ROEBUCK.

"Then," continued the puzzled speaker, "how could he eat all those things in one day?"

The young governess was as much amused at little Edith's mistake as were her two older pupils, but she answered very kindly:

"These things, pet, were intended to feed an immense household—as many people as would fill a village; so you see that it was not really so much as it sounds. And now see what a mistake I have made. I told Clara that deer were not mentioned in the Bible except under other names, and here are 'fallow-deer' as well as 'harts' and 'roebucks.' To think that I should never have seen it before!"

It was Clara's private opinion—in which the others fully joined—that Miss Harson was "just too sweet for anything," and she was obliged now to express this opinion by what the young lady called "her bear's hug." When order was restored, Miss Harson continued:

"There are several verses in which 'hinds' feet' are mentioned as emblems of speed and agility, and we read, 'Then shall the lame man leap as an hart,'* while King David says, 'He maketh my feet like hinds' feet.' There is that beautiful verse in the forty-second psalm: 'As the hart panteth after the water-brooks, so panteth my soul after thee, O God!' Another verse in regard to the hind is found in Jeremiah: 'Yea, the hind also calved in the field, and

* Isa. xxxv. 6.

forsook it because there was no grass.'*
This was part of a prophecy of a terrible
state of things; for the deer always pro-
vides for a retired, protected place for her
fawn, and instead of leaving it she will de-

HART.

fend it with her life. She seems to teach
it also to protect itself by pretending to
be dead, of which I have spoken of before.
A writer who has closely noticed these animals says: 'One day some time ago I was
watching with my glass a red-deer hind
whose proceedings I did not understand
till I saw that she was engaged in licking a

* Jer. xiv. 5.

17

newly-born calf. I walked up to the place,
and as soon as the old deer saw me she
gave the young one a slight tap with her
hoof. The little creature immediately laid
itself down ; and when I came up, I found
it lying with its head flat upon the ground,
its ears closely laid back, and with all the
attempts at concealment that one sees in
animals which have passed an apprentice-
ship of some years to danger, whereas it
had evidently not known the world for
more than an hour, being unable to run or
escape. I lifted up the little creature, being
half inclined to carry it home in order to
rear it. The mother stood at a distance
of two hundred yards, stamping with her
foot.' "

"I s'pose she was scolding about it,
wasn't she?" asked Edith.

"Yes, dear; and when the gentleman
put the little fawn on the ground again, the
deer trotted up to it with evident delight,
and licked it all over again to be quite sure
that her precious child had not been hurt
by the terrible creature who had been hand-
ling it. At any other time she would have

run away from him, but she could not leave her baby."

The children pronounced this "a dear little story," but their governess puzzled them for a few moments by saying that she called it a little-deer story.

"I wonder," said Clara, "if the little fawns care as much for their mothers as the mothers do for them?"

"Only while they are unable to protect themselves, I am afraid," was the reply. "I have read somewhere that a buck of a year old does not even seem to know his mother. A very young one, however, often shows a strong affection, and they are such pretty, graceful little creatures that it is very interesting to watch them. One of our naturalists says again: 'The highest perfection of graceful motion is seen in the fawn of but a month or two old after it has commenced following its mother through the grounds. It is naturally very timid, and is alarmed at the sight of man; and when it sees its dam go boldly up to him and take food from his hand, it manifests both apprehension and surprise, and sometimes something akin to displeasure. I have seen one standing a few rods away face me boldly and stamp his little foot in a fierce and threatening way, as if he would say, "If you hurt my mother, I will avenge the insult on the spot." Ordinarily it will stand with its head elevated to the utmost, its ears erect and projecting somewhat forward, its eyes flashing, and raise one fore foot and suspend it for a few moments, and then trot off and around at a safe distance with a measured pace which is not flight, and with a grace and elasticity which must be seen to be appreciated. A foot is raised from the ground so quickly that you hardly see it ; it seems poised in the air for an instant, and is then so quickly dropped, and again so instantly raised, that you are in doubt whether it even touched the ground ; and if it did, you are sure it would not crush the violet on which it fell. The bound also is exceedingly graceful.'"

Clara was particularly delighted with this description, and all the children appeared to think that they could scarcely be happy again without seeing a little fawn.

DEAR LITTLE DEER.

"How would you like," asked their governess, "to hear about two pet fawns which some one else saw? I know you all like to hear stories."

"Is it a real, true story?" asked Edith, with a happy little wriggle.

"Did *you* see the fawns, Miss Harson?" asked Malcolm.

"Is it one of your own stories, Miss Harson?" added Clara, as though she would not care much about it if it were not.

"'Yes,' to everything," replied the young lady, laughing. "And if you are satisfied, I shall proceed to tell you about them."

Miss Harson then related the story of

NANNY AND BILLY.

The names sound like goats', I know; but the animals were really fawns, and just about as pretty as they could be. They were born among the mountains in Kentucky, and one day, when they were about a month old, a hunter shot the mother-deer, and was just slinging her over his shoulder to carry her away to his cabin when he heard a sort of patter as of little feet, and, looking around, there were two young fawns, the very prettiest he had ever seen, with such soft, bright eyes and glossy coats of light reddish-brown thickly marked with

white spots, while underneath the color
was all white.

"You beauties!" exclaimed the man as
soon as he saw them; and the little creatures stood there looking at him as though
they expected to be carried away with their
mother. He felt sorry now that he had
shot her, although it was done for the sake
of the meat, which he really needed, and,
as his cabin was some distance off, he
could not for a moment think just what
he had better do. It would not be safe to
leave either the dead deer or the live fawns
for a second trip, as the deer would probably be seized by animals or men, and the
fawns, if not captured, would run away;
yet how could he carry all at once? He
plunged his hands into his pockets, scarcely hoping to find a thin, stout rope which
he sometimes carried with him; yet he did
find it, and, tying this securely around the
head of the deer, he dragged the dead
animal along by it and carried the babies
in his arms. His gun he had managed to
fasten on the deer, and some bright animal-
eyes that peered out at him from holes and

bushes as he passed along probably told the brains belonging to them that it was the funniest procession they had ever seen.

It was a tired hunter that reached the lonely cabin as the sun was setting, and even then he had plenty to do. The dead deer had to be attended to, that the flesh might be preserved for eating, and, as for those "little midgets," as the man called the fawns, they were as much trouble as real children. They did not seem to want to run away, but after they had had all the milk they would drink, and had been tucked comfortably into bed on a nice mattress of boughs and leaves, what did they do but get up in the middle of the night, when the hunter was asleep, and go pattering about to explore the cabin? You see, deer are accustomed to trotting about at night in a very dissipated way—it seems to be born with them—and they are, besides, very curious. This inquisitiveness often costs them their lives. The hunter woke from a queer dream about being upset, to find those two little objects pushing at him with their heads to see what he was made

of, and he had to get up and put them
back in their nest, and pile boxes around,
so that they could not get out again.

The next day the man nailed some slats
across the largest of these boxes, in which
the fawns could be comfortably carried, and
he took them away in a wagon to the nearest town.

Just on the outskirts of this town there
was a very pleasant home with large
grounds about it, some parts of which were
almost like real woods, and wide windows
to the house, and two or three verandas.
Ezra Trail, the hunter, knew the place very
well, for he had been there before. It
belonged to Judge Dunleath, and Ezra also
knew that little Miss Cara, the judge's
only daughter, was the very apple of his
eye, and that he refused her nothing. All
that was necessary was to bring Miss Cara
and the fawns together—for the fourteen-
year-old girl was passionately fond of pets
—and the hunter's errand would be accomplished.

"I wonder what Ezra's bringing now?"
said pretty Cara, rather indolently, as she

stood in one of the large windows and watched the grizzled-looking hunter as he "toted" along his box.

Miss Dunleath, Cara's aunt, a dignified lady of middle age, had just looked up from her embroidery at the question, when there was a sudden delighted squeal—for it could not be called anything else—and the excited little girl had bounded out on the veranda. Cunning Ezra had taken the fawns from their prison and put them down near the window. They were too surprised to move, and Cara pounced upon them with "Oh, you darlings!"

At that moment Judge Dunleath appeared in sight, and Cara's frantic "Papa! Oh, papa! Please buy these dear little fawns for me!" speedily resulted in the transfer of a crisp greenback to the hunter's greasy pocketbook, and of the fawns to Cara's rapturous embrace.

"Do not kill them with love, Cara," said her father, laughing, "if they *are* perfect little beauties."

What lovely little pets the fawns were! and how they seemed to grow from day to day! They were very gentle, Nannie especially, while Billy, who was more mischievous, delighted in cutting up the funniest little antics.

But Miss Dunleath did not think Billy altogether lovely, for one day, when she was stroking his back, he reared up suddenly and moved away from her with something like a flash in his eye which quite startled the good lady. Young Mr. Deer did not object to being stroked on his face or head, but he plainly resented the touch of a hand back of his shoulders. Dear little Nannie could be stroked anywhere, and Cara was in the habit of saying that "that child was a great comfort" to her.

They really seemed like children, those little forest-waifs, and every movement was so pretty that it was a pleasure to watch them. They wandered about as they pleased, and the "pat, pat" of tiny hoofs on the veranda usually announced that Nannie was coming to get a drink of water out of her mistress's wash-basin— the only way in which she would take it. Billy was not so particular; when he was thirsty, he would drink wherever he could find water, and, as for eating, it might be charitable to say that he was fond of experiments. One day he was found trying to chew up a small tin plate on which his food had been placed, and not long afterward Judge Dunleath took the chain of his handsome gold repeater from the little scamp's mouth. It had been left on a table where Billy could reach it, and he wished to know if it tasted as good as it looked.

"But he's *so* cunning, papa!" pleaded Cara, hugging the little sinner up close; and then Billy would lick her face and hands and look so pitiful with his great brown eyes!

What good times the little brother and sister seemed to have racing about among the trees and rocks, chasing and tumbling over each other, and never seeming to get hurt in their wildest play! It appeared wonderful that those little slender legs did not break in two when so much scampering was done on them, but the fawns flourished and grew larger and fatter.

One day, though, something very sad happened: poor little Nannie was found lying beneath a ledge of rock with one of her fore legs doubled under her, and she moaned when they tried to move her. Her young mistress was nearly beside herself with grief, but she had a clear little head of her own, and she sent a colored boy at once for the kind doctor who had known her ever since she was a baby, and fortunately he was at home. He seemed as much interested in the little suffering animal as was Cara herself, but he looked rather serious over the poor slender leg that was badly broken and only hanging on by a piece of skin.

"Oh, doctor," said Cara, with quivering lip, "must dear little Nannie die? Can't you tie that poor little leg together somehow?"

"Are you a brave girl?" asked Dr. Still, with a searching glance. "Will you stop crying and help me?—that is, unless you want some one else to care for Nannie. Perhaps your aunt will do it?"

"Yes, indeed, Cara," said Miss Dun-

leath, kindly. "Run into the house, child;
this is no place for you."

"Thank you very much, Aunt Eleanor,"
replied the little girl, with quite a womanly
air, "but Nannie is my pet, you know, and
I should like so much to do everything
myself."

"Go into the house, all of you," said the
doctor, with playful command, "and leave
Cara and me here just to our own two
selves. Let some one bring me out some
thick starch and a long, narrow strip of
cotton-cloth."

Cara wondered if the leg was to be stuck
together with starch, but the doctor did
nothing of the kind. Very tenderly and
carefully he drew the two parts together,
using thin strips of pasteboard for splints,
and then bound them firmly with the long
band dipped in starch and wound round
and round. When this starch dried, the
leg seemed to be encased in a thin board,
and there was no danger of its getting out
of place. It was quite touching to see how
still and good Nannie was through all this
painful work, as if she knew that they were

going to make her poor little leg well again, while Billy stood pondering the matter with an air of great perplexity.

When everything was finished, Cara looked very pale and trembling; but Dr. Still patted her cheek and said,

"You have been such a brave little woman that you are not going to give way now? Nannie is doing splendidly, and there is every reason to suppose that with care her leg will get as well as ever. Whatever possessed the silly, pretty little creature to go and break it?"

Then Billy edged up and concluded to take it out in licking Nannie's head and then Cara's—a practice to which he was much given, but which his little mistress did not enjoy so well as he did. In his exuberant affection he even tried to get at the doctor; but that gentleman threatened to put a pill down his throat if he persisted in running at him with his mouth open, and Billy finally retired.

A cozy nest was made up for the little sufferer in a large basket, which was placed in Miss Cara's room, and she was fed with

the daintiest things imaginable. Dr. Still
looked in upon her once or twice and said
that "his leg was doing well," but Cara
thought it would be a very ridiculous leg
for so large a man.

After a while the little fawn began to go
about again with her injured leg still in the
starched bandage, and it was interesting to
see her managing so well on three legs,
keeping the other one up out of harm's
way. Cara had only to call, "Nan-nie! Oh,
Nannie!" when up the little thing would
trot on her three legs, though not so fast
as she did on four. But she was very playful, and did not seem to suffer much with
it; so there was every prospect of Nannie's
growing up into a large deer. Cara would
have preferred to keep her pet always little,
just as Aunt Eleanor said the other people
at home wanted to keep a certain baby who
would grow into a great girl.

I can scarcely bear to tell you of one
dreadful night when a neighbor stopped
and said that some sheepdogs were chasing the fawns; and when every one turned
out and drove off the malicious brutes, Bil-

ly could not be found, but poor little Nannie was lying at the foot of a tree panting and bleeding and almost dead. There was a dreadful wound in her side, for, with her lame leg, she could not run fast enough to get away from the savage dogs; but Billy had made good his escape.

Again Dr. Still was sent for, and again Cara cried bitterly over her pet; but there was less hope this time, for the doctor said that, although Nannie might recover from the wounds made by the dogs' fangs, she could scarcely get over the terrible fright. Cara held her pet tenderly while the hurt was being dressed, and would let no one do anything for it that she could do herself.

But, in spite of every care, the little fawn never moved again, and died the next day. They buried her just where she was found, and as Cara was stooping over the little mound that covered her lost pet she felt a wet touch on her head and face that startled her, it was so like Nannie. It was only Nannie's brother, who had wandered back again and seemed to want all the love that could possibly be spared for him. He was

18

so lonely, poor little fellow! and Cara declared that if he was only a deer he missed
his playmate and needed an extra share of
petting; so, in spite of losing Nannie, he
seemed to have a very good time, and
played about from morning till night, scampering over every obstacle and often rushing like the wind after nothing in particular.
His michievous pranks, however, did not
always please, and especially his trick of
nibbling at everything to see if it was good
to eat.

One fine day Master Billy set off to explore a field of cabbages, and nipped the
tender white heart from every single head.
This was bad enough for one day, but Billy
did not seem to think it sufficient; for in the
afternoon he rushed at Miss Dunleath with
his head to the ground, quite like a grown-
up deer, and knocked her down on the
gravel-walk. Her arm was considerably
bruised and she was very much frightened.
This led to the discovery that Billy's antlers
were growing, and that he was getting quite
too large and strong for a pet; so a gentleman who lived a few miles off took the fawn

to a large park that he had to keep company with a beautiful greyhound, the two animals, different as they were, soon became the best of friends.

Cara felt quite unhappy at parting with Billy, but she mourned Nannie more, because, as she said, " Billy was alive and safe, but Nannie was gone for ever, and she was such a darling !"

"Do you know, Cara," said the good doctor to her one day, " I think that little fawn has taught you a great deal ? You overcame your own feelings to minister to her sufferings, and showed that you cou d be both firm and courageous. Do not let it stop there, but imitate the example of Him who went about doing good, and learn to be a ministering angel wherever there is suffering to be found in any shape."

When Miss Harson had finished this story, she was almost sorry that she had told it. Clara and Edith were both in tears "for that dear little Nannie," and Malcolm's eyes looked very suspicious as he expressed a frantic desire to own Billy.

"Why, he is a great antlered deer by this time, if he is alive at all," replied his governess, "and not the sort of pet that we should care to have at Elmridge."

"And did you really see them both, Miss Harson?" asked Clara as she tried hard to stop crying.

"Yes, dear, I saw them soon after they arrived, for my mother took me to Judge Dunleath's on a short visit. I was quite a tiny girl then, several years younger than Cara, and I looked at the fawns with great curiosity, but did not care about touching them. Some day, however, I think I must have a pair of my own, and I should not wonder if I kept them loose at Elmridge."

This caused the wildest excitement, which was quenched only by the stern approach of bedtime.

CHAPTER XIV

"NOW," said Miss Harson the next evening, "you all know that the animals we have lately been talking about are useful animals—animals which man has brought into subjection and taught to lighten his labors. We have not yet quite finished with this class, and I should like to have you put on your thinking-caps and see if you can find any more."

"I know one," exclaimed Edith, with great satisfaction. "*Monkeys* are useful, because they go around with hand-organs and get pennies in their caps."

"That is very true, dear," replied her governess, smiling: "they are useful in that way; but no one ever rides monkeys, you see, or makes them carry burdens."

"Is it a very big animal, Miss Harson?"

asked Malcolm, with a look of great intelligence.

"It is not a dwarf, certainly, although
some of the species are larger than others."

"And hasn't he a pretty big name?" continued the young gentleman, who was
quite proud of his discovery. "And doesn't
it sound very much like 'elephant'?"

"It certainly does," said Miss Harson;
"and I suppose that it is because he is such
a great creature altogether it has taken so
much time to get to him. Let us proceed,
then, to become intimately acquainted with
the elephant."

"Why, I thought," said Clara, "that elephants were wild animals, and that people
didn't do anything with 'em but to have
'em in menageries?"

"They are not used in this country,
Clara, for much else, but in the warm
regions where they are found they are
more valuable than even the camel for
some purposes. They are wonderfully
intelligent and interesting animals, and all
the information that we can get about
them is fairly honeycombed with stories."

This was a delightful prospect, and the little Kyles hoped that in proportion the "talk" would be as large as the size of the animal.

"It cannot possibly be a small one," was the laughing reply, "and, although the elephant does not properly belong with donkeys and mules and camels and reindeer, I have thought it better for my purpose to put the animals into different classes from those in which naturalists put them. We have, therefore, had little neighbors and home-animals, and now we have useful animals. Not that home-animals are not useful, for many of them are eminently so; but the present is a class of animals that we are not accustomed to see in use."

"People ride on elephants in pictures," said Clara.

"So they do in menageries," replied her brother, "for you know that papa took me once, a great while ago, when I was a little boy. I was frightened and cried to come home, but I wouldn't do that now."

"No, indeed!" said little Edith. "Why,

ELEPHANT-RIDING.

I wouldn't cry now.—Would I, Miss Harson?"

"I think not, dear ; and if papa does not object, I've a notion of trying you quite soon."

When was papa ever known to object to anything that Miss Harson proposed ? So the children considered their going to the menagerie as good as settled, and were joyful accordingly.

"We must begin upon our very large animal," continued the young lady, "if we have any hope of getting through with him ; and you will probably be surprised to hear that there are two kinds of elephants—the Asiatic, or Indian, and the African. They are exactly alike, however, in their habits, and very much so in appearance ; but look closely at these two pictures and tell me what differences you can find."

"The head of the Indian elephant is smaller," said Malcolm, "and so are the ears."

"Yes," replied his governess ; "the ears of this African elephant are something wonderful to behold. See how they lie back and hang down just above his fore legs—

as some one has said, like immense leaves
or a great fan on each side."

"That is just what they do look like," exclaimed Clara. "Miss Harson, look at these
little hills on the African elephant's back;
the other one doesn't have 'em."

"There are two decided waves there,
while the Indian elephant's back is more
like one large hill."

"What little bits of eyes they've got!"
said Edith, who was anxious to make some
discoveries of her own. "I shouldn't think
they could see much."

"The eyes are smaller in proportion to
the elephants' size," said Miss Harson,
"than are those of any other animal, and
their sight is not very good. It is thought
by naturalists that elephants' eyes are small
for protecting them in the dense thickets
among which they love to roam, and they
are also furnished with a sort of membrane,
which can be drawn down at pleasure, as
a further protection against small twigs
and thorns. You will notice, Malcolm, that
the head of the Indian elephant is not only
smaller, but is differently shaped, being nar-

rower. The head of an elephant is the strongest part of the animal, and the thick bone will even flatten a bullet that is fired against it."

"I suppose, Miss Harson," said Clara, "that the Indian elephants live in India?"

"Yes, in India, with its provinces and adjacent islands, especially in the large island of Ceylon. They really inhabit the whole of Southern Asia, and are therefore called Asiatic elephants. The African elephant is found on the whole western side of Africa, as far down as the Cape of Good Hope."

"I think," said Malcolm, "that the queerest thing about an elephant is its trunk. It looks like an immense nose curled up at the end, and yet it is like a hand, because its owner can take things up with it and hold 'em."

"It is a nose," was the reply, "and it is also a hand, because it has a sort of finger and thumb. Between these are the openings of the nostrils. There is a great deal to learn about this trunk, but first we will try to get some general idea of this whale among land-animals. As for its size, an ordinary elephant is eight or nine feet high and about fifteen feet long. African elephants are sometimes twelve, and even fourteen, feet in height, and the weight of such a huge creature is six to seven thousand pounds."

"Wouldn't it be dreadful," said Edith, "to have an elephant tumble over on any one?"

"Dreadful indeed, Edie; but fortunately there are none of these monsters roaming about in the woods at Elmridge. When elephants are wickedly disposed—as they sometimes are, even those which have been tamed—they make use of this terrible weight to crush a fancied enemy by driving him against a wall. It makes them powerful engines of war, for which they have often been used, as there is no escape from the close approach of such an immense moving mass."

"Miss Harson," asked Clara, "is a real elephant the color of that dark-gray canton flannel one that you made for Edith ever so long ago?"

"Something of that color," replied the young lady, "but darker and duller-looking. I should say that elephants are of a lead-color, though sometimes they are nearly black. Yet white elephants are frequently found—or those that are called white. These are never entirely white, but very light-colored. They are usually Indian elephants, and are owned only by royalty. One of the titles of the king of Siam is king of the white elephant. And now what next?"

"These queer things sticking out by his trunk," said Edith; "what does he do with 'em, Miss Harson?"

"Kills his enemies sometimes, dear, for these are tusks, or teeth; and you remember my telling you about the dangerous tusks of the wild boar? Those of an ordinary Indian elephant will weigh as much as sixty pounds, and those of the African species over one hundred. The latter are from six to eight feet long, and they are the most valuable part of the elephant because of the numerous uses for the ivory of which they are formed."

"What queer-looking feet he's got," said Malcolm—"without any ankles!"

"His legs do seem to come to a very sudden end," replied Miss Harson; "but if you look closely, you will see that the 'queer-looking feet' are broad hoofs with five nails, although the hind feet seldom have more than four. Sometimes there are a great many more of these nails, which vary greatly in number. The sole of this hoof-like foot is nearly round, and in an elephant of common size it will measure about twelve inches across. The huge body is supported upon very solid pillars, and the pillars themselves have most substantial foundations."

"I should think so!" exclaimed Clara. "How I'd hate to be an elephant!— Wouldn't you, Miss Harson?"

"On account of having such big feet?" laughed her governess. "I have not considered the advantages of being an elephant, but, whatever they may be, there is one thing, Clara, in which the most insignificant human being is superior to the grandest animal that ever lived."

"I know," said Clara, softly. "You mean the soul, Miss Harson?"

"Yes, dear Clara—the soul with its blessed hope of immortality through Jesus Christ our Lord. When lost in wonder over the strength and power and marvelous formation of such a huge animal, we can remember that after its allotted years on earth these creatures all perish and are as though they had never been, while the helpless infant whom it could crush with one touch of its foot dies only to live for ever."

"How long do elephants live?" asked Malcolm, presently."

"One hundred years," replied his governess, "seems to be the ordinary period of an elephant's existence, but some live much longer, as great an age as three hundred years having been known. If such a powerful creature, living so many years, were a meat-eater or a carnivorous animal, he would destroy all other living things, but he feeds entirely upon vegetables, with a strong partiality for the twigs of trees."

"But where does he put anything to

eat?" said Edith, puzzling over the picture.
" He hasn't got any mouth."

"It does not appear there, dear, but here
is a picture which will show you a mouth
directly under the trunk."

"All the same, ma'am," said Malcolm, "I
don't see how he ever gets anything into it.
And how does he get his trunk up into the
air like that?"

"With the help of its forty thousand
muscles. 'We need not, therefore, be surprised if this instrument be strong enough
to tear up a tree and delicate enough to
seize a pin. There is no animal structure
in the least like the trunk of an elephant,
but, though the mechanism is unique, it is
altogether complete for its purposes.' This
wonderful trunk has been called 'the elephant's hand' and 'the snake-hand,' and
the Caffre of Africa, when he has killed an
elephant, still has a feeling of superstitious
awe for the trunk, which he cuts off and
buries, saying, 'The elephant is a great
lord, and the trunk is its hand.' This member has neither bone nor cartilage, yet it
can be contracted, moved up or down, to

the right or the left, and shot out, when necessary, from a foot to five feet long. The centre of the trunk is pierced by two long canals which are the continuation of the nostrils, and they are separated by a fatty substance which is less than half an inch thick. When these channels reach the centre of the bone in which the tusks are planted, they suddenly turn toward the outer part of this bone with a semicircular curve, and they are so much narrower here that unless the elephant uses its muscles to dilate them they act as valves to keep from going higher any liquid that may be taken into the trunk. The canals widen again beyond this point and curve back; so that the elephant can use his trunk as a vessel for holding water, to be resorted to as he has need of it. Now," added the young lady, "I know that this is not nearly so interesting as a great many other things about the elephant—I do not enjoy it so much myself—but it is important for us to understand how this mysterious trunk can be put to so many different uses. There is a great deal of learned writing about the

19

two sets of muscles by which the trunk is
moved, which muscles are called longitudinal and transverse.—Can you tell me what
that means, Malcolm?"

"Isn't 'transverse' 'going around,' and
'longitudinal' isn't?"

"'Transverse' is slanting," was the smiling reply; "you did not present your ideas
in a scientific form. But this will do for
muscles at present. You see in these ends
of the trunks the finger and thumb, and in
these the muscles are even more flexible,
so that the animal can seize such objects as
he wants with his finger and hold them fast
with the help of the thumb. The trunk of
the elephant may be first regarded as an
instrument for collecting his food. He feeds
upon all vegetable substances, from the
leaves of trees and the coarsest grass to
the most farinaceous grain and the choicest
fruit. Though his enormous bulk—requiring that his provender shall be in large
quantity—renders a plentiful supply of the
commoner vegetable productions necessary to him, yet his palate is pleased with
delicacies. For this reason the strength

and the minute touch of his proboscis are equally available in the collection of his daily supplies. If he meet with long herbage, he twists his trunk spirally round the roots and crops them off. The bundle which he gathers is then held between what we have called the finger and thumb of the trunk, and is thus conveyed to the mouth. If the objects which he is collecting are too small to repay him for the trouble of carrying them to his mouth, he holds them one by one behind his thumb till he has gathered enough for a load. Thus, if he find a small root, he seldom eats it at once, but collects two or three, holding each as you see in the book in Clara's hand. When the object which he wants requires force for its removal or is difficult to reach, he completely curls his trunk around it, and in this way, elevating himself upon his hind legs, he pulls down the tall branches of the trees of the forests which are his natural domain."

The children pressed forward with great interest to see the pictures of the elephant's trunk in different positions, and of the huge

animal itself breaking down the branch of
a tree.

"It looks like an enormous pig," said
Clara.

"It has some right to that look," was the
reply, "for it is a relation of our grunting
friend."

The more they looked at the picture,
the stronger the resemblance seemed to
grow; and, remembering the wild boar's
tusks, the children wondered that they
had not thought of it before. But Miss
Harson did not wonder in the least, as it
would be quite out of the usual fashion to
associate elephants with pigs.

"I should think," said Malcolm, "that it
would be very hard work for the elephant
to get the end of his trunk into his mouth.
How does he manage it, Miss Harson?"

"By means of those forty thousand
muscles, Malcolm. He cannot get his
mouth to the food, so he carries the food
to his mouth by doubling his trunk into a
sort of loop, the end of which reaches his
mouth. 'As an organ of touch,' says a
naturalist, 'the proboscis of the elephant is

exquisitely fine. Elephants sometimes go
blind, and under that privation the poor
animal can not only collect its food and
discriminate as to its quality by this wonderful instrument, but can travel without
much difficulty over uneven ground, avoiding lumps and hollows and stepping over
ditches. The creature, in such circumstances, rarely touches the ground with its
trunk, but, projecting it forward as far as
possible, lets the finger—which is curled
inward, to protect the nostrils—skim along
the surface.' "

"Doesn't an elephant's trunk sometimes
get hurt?" said Clara.

"Very seldom, I think," replied her
governess, "because the elephant takes
such good care of it. If threatened with
danger, his trunk appears to be his first
thought, and he raises it up as high as
possible to get it out of the way of harm.
'If this delicate organ be in the slightest
degree injured, the elephant becomes wild
with rage and terror. He is even afraid
of a dead tiger, and carefully moves his
trunk out of reach. The care with which

he endeavors to put his trunk beyond danger
makes him extremely cautious of using it
as a weapon. He rarely strikes with it,
though he will frequently throw clods and
stones with it at objects which he dis-
likes.' "

"I suppose he uses his tusks instead,"
said Malcolm; "they must be equal to
two swords."

"Yes; some naturalists call these formidable weapons the elephant's defences,
and they enable him not only to clear his
way through the thick forests in which he
lives by rooting up small trees and tearing
down cross-branches—in doing which service the tusks effectually protect his face
and proboscis from injury—but they qual-
ify him for warding off the attacks of the
wily tiger and the furious rhinoceros, often
securing him the victory by one hard blow,
which transfixes the assailant to the earth.
These tusks are said to correspond, like
those of the boar, with the canine, or eye,
teeth in other animals; but, unlike those
of other animals, they are formed of that
smooth, beautiful substance which we know

as ivory. Some of these tusks are too heavy for a man to lift, and the largest ever known weighed three hundred and fifty pounds."

" Didn't they make the elephant tired?" asked Edith, pitying the great animal for having to carry such heavy things around.

" No, Edie," was the reply; " these heavy tusks were probably no more for the immense elephant that owned them to carry than the boar's tusks are for him. In some of these pictures you will see that the tusks grow directly downward like the trunk, and this gives the animal a very singular expression. These weapons are often used to turn up the ground to get at roots and bulbs, in which the elephant delights, and which he discovers before they are detected by his fine sense of smell. Where wild elephants are found whole acres will often be seen ploughed up in this way."

CHAPTER XV

"MISS HARSON," asked Clara, "what do people do with ivory?"

"Not nearly so many things as they formerly did," was the reply. "Years ago, when everything from the East Indies was much sought after, ivory was in great demand, and many beautifully-carved articles in ivory were brought from over the sea. Exquisite boxes and book-covers were made of it, also chessmen and card-cases, and even larger articles, such as cabinets and statues. It was also used extensively in inlaid work, and ebony inlaid with ivory was considered very beautiful."

"Isn't it used for piano-keys?" said Malcolm. "They always remind me of immense rows of teeth."

296

"Yes; that is one of its principal uses at the present day, and piano-keys are not unlike great flat teeth.   In the Bible there is no mention made of elephants, although so many other animals are spoken of, but ivory, the product of elephants, is frequently alluded to.   The first mention of it is found in First Kings: 'Moreover, the king made a great throne of ivory, and overlaid it with the best gold.'*   This is part of the description of Solomon's magnificent palace, 'of which this celebrated throne with the six steps, and the twelve lions on the steps, was the central and most magnificent object.'   Solomon was the richest king that ever lived, and in his time ivory was so valuable that it was ranked among the wonders to be seen in his palace.—Read the twenty-second verse of this same chapter, Clara."

"'For the king had at sea a navy of Tharsish with the navy of Hiram: once in three years came the navy of Tharsish, bringing gold, and silver, ivory, and apes, and peacocks.'"

* 1 Kings x. 18.

"Now, Malcolm," continued Miss Harson, "let us hear what the prophet Ezekiel says of ivory in speaking of the greatness of Tyre.   Chapter twenty-seven, verse fifteen.'"

Malcolm read with some surprise:

"'The men of Dedan were thy merchants: many isles were the merchandise of thine hand: they brought thee for a present horns of ivory and ebony.'   'Horns'?" he repeated.   "Why, Miss Harson, I thought you told us that tusks were teeth?"

"Yes," replied his governess, "and the Hebrew word for 'ivory' means 'a tooth,' the Old Testament having been, as I have told you, first written in Hebrew.   The verse which you have read does not really mean that ivory is made of horn, but of 'uncut tusks.'"

"Is there any more about ivory in the Bible, Miss Harson?" asked Edith.

"Yes, dear; in the Psalms it is written: "'All thy garments smell of myrrh, and aloes, and cassia, out of the ivory palaces, whereby they have made thee glad.'*   It is

* Ps. xlv. 8.

thought that 'ivory palaces' meant boxes or chests inlaid with ivory in which the royal garments were laid with perfumes.   There are other verses, though, in which houses really are meant; and in First Kings we read of the 'ivory house'* that was made by wicked King Ahab—not that the whole house was built of ivory, but only that it was inlaid with it. 'Houses of ivory' are mentioned by the prophet Amos.† 'Woe to them,' says that prophet, 'that lie upon beds of ivory, and stretch themselves upon their couches!' In those days, and long afterward, the beds of the wealthy were adorned with ivory. In the New Testament ivory is mentioned only once, and that is in Revelation: 'And the merchants of the earth shall weep and mourn over her, for no man buyeth their merchandise any more: the merchandise of gold, and silver, and precious stones, and of pearls, and fine linen, and purple, and silk, and scarlet, and all thy fine wood, and all manner vessels of ivory, and all manner vessels of most precious wood, and of brass, and

* I Kings xxii. 39.   † Amos iii. 15.

iron, and marble.'* This, like the description of Solomon's throne, shows that ivory was counted in those days among the most valuable things."

"Miss Harson," said Clara, presently, "I don't see how such creatures as elephants could ever be caught to get the ivory. I suppose that their tusks don't drop off, do they?"

"No, indeed; they are too firmly fastened in the thick bone of the head for that. It is no easy matter to get them out even after the animal is dead. The wild elephant is a formidable creature which not even the fiercest beast can attack with impunity, and ivory-hunters have frequently lost their lives in attempting to kill him. He is, besides, seldom seen alone, as these great animals seem to enjoy one another's society and roam about in herds. A traveler who has watched them 'in the deep solitudes of a tropical wilderness' says 'that a herd of elephants browsing in majestic tranquillity amid the wild magnificence of an African landscape is a very noble sight.' Follow-

* Rev. xviii. 11, 12.

ing the footprints of a band of these
gigantic creatures, the same traveler con-
tinues: 'It was in the groves and jungles
that they had left the most striking proofs
of their recent presence and peculiar habits.
In many places paths had been trodden
through the midst of dense thorny forests
otherwise impenetrable. They appeared to
have opened these paths with great judgment, always taking the best and shortest
cut to the next open savanna or ford of the
river, and in this way they were of the
greatest use to us by pioneering our route
through a most difficult and intricate country
never yet traversed by a wheel-carriage, and
a great part of it, indeed, inaccessible even
on horseback except for the aid of these
powerful and sagacious animals. In such
places the great bull-elephant always
marches in the van, bursting through the
jungle as a bullock would through a field
of hops, treading down the thorny brush-
wood and breaking off with his proboscis
the larger branches that obstruct his passage; the females and younger part of the
herd follow in his wake in single file; and

in this manner a path is cleared through the densest woods and forests such as it would give the pioneers of an army no small labor to accomplish.' This traveler saw a great number of trees which the elephants had torn out of the ground and turned upside down that they might feast at their leisure on the soft, juicy roots of which they are fond. With a very large tree the elephant would use one of his tusks as a crowbar, pushing it into the ground under the roots, so that he could easily pull the tree out with his trunk.'"

"No wonder," said Malcolm, "that elephants can do so many tricks after they are taught, if they can do all that without being taught."

"An elephant can do a great deal else without being taught, and among his natural accomplishments is that of cooling himself off by drawing from his throat, with the help of his trunk, a supply of saliva, which he frequently showers all over his skin. He also takes up dust and blows it over his back and sides to keep off the flies, and he often fans himself with a great bough, which he uses as handily as possible. A poet has written of some of these habits of the elephant:

> " ' Trampling his path through wood and brake,
>   And canes which crackling fall before his way,
>   And tassel-grass whose silvery feathers play
>   O'ertopping the young trees,
>   On comes the elephant to slake
>   His thirst at noon in yon pellucid spring.
>   Lo! from his trunk upturned aloft he flings
>   The grateful shower, and now
>   Plucking the broad-leav'd bough
>   Of yonder plume, with waving motion slow
>   Fanning the languid air,
>   He waves it to and fro.' "

It seemed wonderfully comical to think of elephants fanning themselves of their own accord. It was much more amusing than the same performance after they had been trained to do it.

" Perhaps, then," said Miss Harson, rather mischievously, " you will not care to hear about any trained elephants?"

This sounded almost cruel, and the speaker declared that she was quite overwhelmed by such surprised and disappointed looks.

"We shall come to them before long," she hastened to add, " but just now you wish to know how these huge animals are captured. They are hunted for different reasons: the native African wants their flesh to cook and eat, also the tusks, which he is often obliged to get for the petty king or chief to whose tribe he belongs; and the white man hunts them sometimes just for the excitement and glory of killing such large and powerful creatures, and at other times he wishes to take them alive to be tamed and perhaps added to some collection of animals exhibited in other countries. 'When we consider the enormous strength of the elephant, which enables him to break through all ordinary means of confinement, with his ability to resist any violent attack and with sagacity to elude any common stratagem, it is evident that the business of his capture must be a task requiring equal courage and activity, with great skill and presence of mind in the individuals engaged in it.' In Africa, and also in Ceylon, an elephant is often taken by the natives in a pit which is covered with a slight platform of grass and branches, and a tame elephant is often used to lead a herd of these ani-

CAPTURING AN ELEPHANT.

20

mals toward it. The leader of the herd no
sooner touches the unsafe spot than down
he goes, and the others run away in terror.
The one secured is kept in the pit until he
is quite subdued and it seems safe to bring
him out ; then large bundles of jungle-grass
are tied up and thrown to him, and with
these helps he reaches the top by degrees.
' The elephant will do the same if he is
swamped in boggy ground, thrusting the
bundles of grass and straw into the yielding earth with his heavy feet and placing
them so around him with his trunk that he
at last obtains a firm footing.' In Northern
India, where the elephants are comparative-
ly small, they are often captured with a slip-
knot, somewhat after the fashion of catching
wild horses with the lasso. ' The hunter,
seated on a docile elephant round whose
body the cord is fastened, singles out one
from the wild herd, and, cautiously approaching, throws his pliable rope in such
a manner that it rests behind the ears and
over the brows of the animal pursued. He
instantly curls up his trunk, making an effort to remove the rope, which with great

adroitness on the part of the hunter is then passed forward over the neck. Another hunter next comes up, who repeats the process; and thus the creature is held by the two tame elephants to whom the cords are attached till his strength is exhausted.'"

"It seems mean of the tame elephants," said Malcolm, but Clara and Edith thought it was "nice," because it looked as if the tame ones were so contented.

"I think you are both right," replied their governess, "and the account shows how thoroughly obedient these great animals can be made.—The most valuable elephants are those of the greatest size and strength, and these are usually ferocious and wander about by themselves or in little companies of twos and threes. They do great damage to farms and gardens, and seem bent on destroying everything they encounter. Such an animal is sometimes followed for several days and nights before he is caught. The catching is accomplished by the help of two or four trained female elephants, called *koomkies*. 'The females gradually move toward him, apparently unconscious of his presence, grazing

with great complacency, as if they were,
like him, inhabitants of the wild forest. It
is soon perceived by them whether or not he
is likely to be entrapped by their arts.   The
drivers remain concealed at a little distance
while the *koomkies* press round the unhappy
goondah, as this sort of elephant is called.
If he suffers himself to be cajoled by his
new companions, his capture is almost certain. The hunters cautiously creep under him,
and while he is thus amused they fasten his
fore legs with a strong rope. It is said that
the wily females not only will divert his attention from their *mahouts*, or drivers, but
will absolutely assist them in fastening the
cords.' "

This was delightfully like a story, it
seemed so hard to believe, and Miss Harson read to her young listeners from the
Sabbath-School Visitor a description of another mode of taming elephants:

" 'A large party of men—more than
three hundred—go out together into the
jungle; that is, the wild country inhabited
only by wild beasts. They take a number
of tame elephants along. Scouts are sent

on ahead, who soon find a herd of perhaps
forty or fifty wild ones. Then the party
of hunters separate in two diverging lines
and slip off into the forest, leaving two
men behind in their path to stand guard,
one in every fifty yards. When the two
lines have passed quite behind the elephant-
herd, they come together again, making a
circle sometimes two or three miles around.
In the midst of this the wild elephants are
caught, though they do not at once find it
out. It does not take a long time for the
men to put up a light fence of split bamboo
all around the circle. Fires are kindled for
them to cook their rice by, and also to
frighten the elephants. If one of the elephants rushes toward the fence, it is the
duty of the nearest guards to thrust into
the fire long bamboo poles, which send up
showers of sparks and explode with a loud
crack like a pistol; this frightens the elephant, and it runs back. Night comes on,
and the animals make strange trumpeting
noises through the forest; the fires light
up the tree-branches, making a weird
scene. The next day the men build a

strong place like a pound, and after a while they manage to drive either a part or the whole of the herd into it, and shut the door. The creatures are excited and angry

TAMING ELEPHANTS.

at being made fast. Now appears the great usefulness of the tame elephants: no hunter would dare venture into the pound except upon the back of one of these gentle creatures. The wild ones look up in surprise to see one of their own kind behaving so submissively. The hunters contrive to slip a rope around the legs of one of their captives, and, the tame elephant helping, they push and drag him out of the pound. He is tied to a tree-stump, as is shown in the picture, when at first he strikes with his trunk at the keeper and behaves very savagely; but when he sees that the tame elephant standing near him for an example never kicks nor strikes, he will begin to grow more quiet, and in two or three days, usually, he will become so docile as to take sugar-cane from the hand of his keeper.'"

"The men who are tying him look so little," said Clara, when the story was finished, "among these big animals! How easily they could tread on 'em and kill 'em!"

"Yes," was the reply; "that is the wonderful part of it, and it shows how inferior the largest and most intelligent animal is to an ordinary man because God has willed it so. In the beginning he gave man the dominion 'over every living thing that moveth upon the earth,' and sooner or later he obtains it. In one of the sacred books of the Hindus it is written, 'The mind is stronger than an elephant, whom men have found means to subdue, though they have never been able to subdue their own inclinations.'"

"This superiority of human mind over the immense physical bigness of the animal," continued Miss Harson, "is shown in the way hunters contrive to capture this great creature. They resort to all manner of ingenious devices to entrap him; of some of these we have already learned. There is another way in which the hunter is said to deceive the elephant. He pretends to be running away, and entices the huge animal into chasing him down a steep hill. When the elephant is in motion, the hunter quickly turns and runs up hill again. His pursuer, by reason of his vast size, is unable to check himself at once, and the hunter has him in his power. Thus mind in the hunter is more than a match for bodily strength in the elephant."

ELEPHANT RUNNING DOWN HILL.

"But how," asked Malcolm, "do they manage to tame an elephant after they get him caught? I should think he'd be so mad that he'd smash up everything."

"He probably would if he had the opportunity, but this is well guarded against, and he is tamed by degrees. 'The animal is carefully attended upon; all his necessities are diligently supplied; he has abundance of food and drink; his skin is kept cool by continued applications of water; the flies that irritate him are driven off. One man—his intended keeper—is always about him, soothing him by the most diligent kindness. The animal gradually learns that his comforts must depend upon the will of his keeper, and he allows him, therefore, to approach him, and at length to get upon his back. As the elephant gains confidence the keeper is more bold, and soon takes his position upon the neck with the iron hook, ready to direct him by catching hold of his ear or pressing it into his skin. To this rough monitor he gradually yields entire submission, as the horse submits to be urged on by the spur. It is generally as long as six months before the elephant is rendered perfectly obedient to his keeper, so as to be conducted from place to place without difficulty. Once tamed, there seems to be no limit to his obedience and capacity. There is, of course, a great difference in elephants in this respect, as some are naturally more gentle than others, and one with a furious temper will break out upon the slightest provocation, and will even kill the object of its rage. But they are often very affectionate under kind treatment, and an elephant once became so fond of a child that he would not take his food unless his little friend was present; and when the child slept, he was constantly busy driving away the flies.' "

"I suppose," said Clara, "that he flapped his big ears at 'em if he was an African elephant."

"I do not know, dear," was the laughing reply, "but it seems probable that he either did that or waved his trunk to and fro. I only wonder that such a clumsy nurse did not contrive to kill the child, but it seems that in India the wife of a *mahout*—you know what that is now—has been known to give her baby in charge to an elephant while she went away on some business, and the animal's care and sagacity excited the amazement of an officer who watched him. 'The child—which, like most children, did not like to lie still in one position —as soon as left to itself would begin crawling about, in which exercise it would probably get among the legs of the animal or become entangled in the branches of the tree on which he was feeding, when the elephant would in the most tender manner disengage his charge either by lifting it out of the way with his trunk or by removing the impediments to its free progress. If the child had crawled to such a distance as to verge upon the limits of his range—for the animal was chained by the leg to a peg driven into the ground—he would stretch out his trunk and lift it back as gently as possible to the spot whence it had started.'"

This pleased the little Kyles so much that Miss Harson read a short article written by a missionary in India:

ELEPHANT EATING SUGAR.

"'Once, while on a visit to a friend, we were taken to a deserted city to spend the day. Elephants were sent us on which to make a portion of our journey; when we reached our destination, the elephants were turned into an enclosure. Mosquitoes were abundant, and troubled the huge creatures, though they soon managed to rid themselves of their unwelcome visitors. The earth in the enclosure had been trodden into fine dust; with this one of the elephants filled his trunk, and, holding it aloft, showered the dust over his body, repeating the process until his body was thickly coated. The other elephants followed his example.

"'I have heard of one tame elephant whose doings were quite wonderful—such as taking a glass of sugar from his keeper's hand, eating the sugar and giving back the glass, as we see him doing in the picture. The keeper of this elephant was given each day a certain amount of flour, with which he was expected to make cakes of bread for his charge. For a time the man honestly gave to the elephant the full

amount, but at length he began to lay aside
for himself one or two cakes, and gradually
increased the number. The elephant seemed
to take no notice, but one day, as the cakes
were brought to him, he turned over the
entire pile with his trunk, as if counting
them; then, rushing toward the astonished
keeper, he wound his trunk about the man's
body, and, taking him to an open well broad
and deep, he held him above it, shaking him,
and seeming as if he would really drop
him into the depths below, but at length
releasing him. You may be sure that the
keeper took no more cakes belonging to
that elephant.' "

The children seemed more and more delighted with each story, and Miss Harson
said that she must tell them about an elephant that made an express-wagon of himself.

"This sagacious animal," continued the
young lady, "would swim across the Gan-
ges loaded with parcels, and would then
unload himself without the least assistance.
Another one—also an Indian elephant—was
seen early one morning marching alone

into the courtyard of the fort at Travancore with a heavy box on his trunk. Having put this down and departed, he soon appeared again with a second box, which he deposited by the side of the other one. Again and again he came with the same burden, until there was quite a pile of boxes arranged in the most orderly manner. These boxes were filled with money and jewels belonging to the rajah of Travancore, who had died in the night, and they were removed in this singular manner to the fort for greater safety."

"Only think," said Malcolm, "how it would seem to get a parcel or a box by elephant instead of by express! Wouldn't it be funny, though?" His sisters seemed to think that, although it might be funny, they would much prefer the ordinary expressman.

"I shouldn't think an elephant would carry things for people," said Clara, "when he is sent off without any driver. He might run away then."

"The tame elephants do not seem to care to run away," was the reply, "and there is

a queer legend about the way in which they
first came to enter into the service of man-
kind. The story says that when this happened there were not nearly so many elephants in the world as there are now, and
that they all lived together. Their hind
legs, which now bend forward, like the legs
of a human being, then bent backward,
like the legs of a quadruped—as the story
says. The people living in those days had
none of our useful animals to help them
carry things around, and they had no wag-
ons or carts to put anything in. This made
it very hard for them, and at last they began to think that the elephants, who were
so big and strong, might just as well help
them; so some of them went to the leader
of the great herd and talked to him about
it. They promised to supply the elephants
with the fruit and vegetables which these
animals particularly liked from their gardens if they would agree in return to carry
the people and their heavy things whenever they desired it. The head-elephant
was greatly pleased with the offer, but he
said that he did not see how the plan could

21

be carried out. They were so high that they would have to kneel to let the people get on their backs, and it was so hard for them to get up again when they once got down that it would be quite impossible to do it at all with a heavy load to lift. This was reasonable, but very discouraging, and the men went back to their companions. Some others then visited a very wise witch in the neighborhood and asked her advice. Having received a handsome present in payment, she went at night where the elephants were all lying asleep on the ground or leaning against trees, and without their knowing anything about it she managed to make their legs all bend inward. In the morning the head-elephant first discovered this change in himself, to his great surprise, and the whole herd found out by degrees that they could rise up quite easily. Then the men came to them again and persuaded them to make the agreement they had proposed; for when it was seen that an elephant could really get up and shuffle off with a great load of things on his back, there was nothing more to be said. It is reported, though, that a great many of these animals became very much dissatisfied with the heavy work they had to do for men from morning till night, and so they wandered off from the herd and settled in Asia and Africa and the other places where they are found, and before they will do any labor now they have first to be caught and then tamed."

"Isn't that true—what you have just been telling us, Miss Harson?" asked Edie, with great interest.

"Why, no, dear," replied her governess, smiling; "how could it be true, when we know that animals do not talk? Elephants are very fond of melons and rice and tender leaves, and all the delicacies that were promised them in the legend; and, considering their great strength and intelligence, it is quite probable that if they could have made such an arrangement—and if their hind legs had ever been bent the other way and could have been bent forward, as the story says—they would have made it."

"Well," said the little girl, contentedly, "then it might be true."

"There is an amusing old story," continued Miss Harson, "of some mischievous Eastern tailors and an elephant. The elephant was on his way to the river under charge of his master. Passing a shop where tailors were at work, he put out his trunk with the hope of receiving a gift of fruit; in place of this, one of the tailors stuck a needle into the extended proboscis, thinking it a good piece of fun. The elephant quietly went to the river, and after drinking filled his trunk with muddy water. When, on his way back, he was again at the tailors' shop, he again put in his trunk, but it was to deluge the cruel men with water. It would be well for those who are fond of playing tricks on others," said the governess, "to ask themselves how they like it when the tables are turned and the tricks are played on them. 'To do to others as we would that they should do to us' is a good rule for tailors, and everybody else."

Malcolm thought the shower-bath which the tailors received served them right for their unkind treatment of the poor elephant.

CHAPTER XVI.

HARD WORK.

"MISS HARSON," asked Malcolm, "didn't people formerly use elephants in fighting battles?"

"Yes," replied his governess; "these animals were used in very ancient times both to carry soldiers into battle and to attack the enemy themselves. Besides trampling and crushing those within reach of their powerful feet, these war-elephants have been known to stretch out their trunks and pick up soldiers, whom they placed in the hands of their riders. Alexander the Great was obliged to contend with the elephants employed by the Indian monarch whose kingdom he invaded, and after gaining the victory he used these powerful animals in his own service. Elephants are particularly associated with India, and the

early princes of that country depended
largely upon them for their success in war.
Whoever could muster the largest number
of trained elephants was sure of victory.
A very small army with a single elephant
has been known to disperse a much larger
force, which fled at once because of the terror
which the animal inspired; but an Indian
emperor named Baber, who seldom used
elephants in war, on going to meet a ter-.
rible adversary, spoke of nerving himself
for the encounter, as a Christian might, in
these beautiful words: 'I placed my foot
in the stirrup of resolution and my hand
on the reins of confidence in God, and
marched against Sultan Ibrahim, the son
of Sultan Iskander, the son of Sultan
Behlûl Lodi Afghan, in whose possession
the throne of Delhi and the dominions of
Hindustan at that time were, whose army
in the field was said to amount to a hundred
thousand men, and who, including those of
his amirs, had nearly a thousand elephants.' "

It seemed impossible to take in the idea
of such a number of huge creatures, and

Edith asked, in great perplexity, where they all stayed.

"I suppose, dear," was the reply, "that they had stalls of some kind, but possibly only a thick stake driven into the ground with a chain attached to fasten them to. In the warm climate where these animals belong they do not need the shelter which has to be provided for them in cold regions."

"Do the people in India use elephants now in fighting?" asked Malcolm.

"Only in some very remote provinces, for India, you know, now belongs entirely to England. But even in modern times elephants are made very useful in war, although not taken into battle. They have assisted in dragging heavy cannon, pushing the carriage-wheels with their heads and trunks, and immense guns have been carried on their backs. Here is an interesting account of elephants carrying some guns up a hill which shows not only their usefulness, but also their great caution and intelligence :

"'Having cut a good deal of the most

prominent part of the hill away and lain trees on the ascent as a footing for the elephants, these animals were made to approach it, which the first did with some reluctance and fear. He looked up, shook his head, and when forced by his driver roared piteously. There can be no question, in my opinion, that this sagacious animal was competent instinctively to judge of the practicability of the artificial flight of steps thus constructed, for the moment some little alteration had been made he seemed willing to approach. He then commenced his examination and scrutiny by pressing with his trunk the trees that had been thrown across, and after this he put his fore leg on with great caution, raising the fore part of his body so as to throw its weight on the tree. This done, he seemed satisfied as to its stability. The next step for him to ascend by, which we could not remove, was a projecting rock. Here the same sagacious examinations took place, the elephant keeping his flat side close to the side of the bank and leaning against it. The next step was

against a tree, but this, on the first pressure
of his trunk, he did not like. Here his
driver made use of the most endearing
epithets, such as, "Wonderful, my life!"
"Well done, my dear!" "My dove!"
"My son!" "My wife!" but all these affectionate appellations, of which elephants
are so fond, would not induce him to try
again. Force was at length resorted to,
and the elephant roared terrifically, but
would not move. Something was then
removed; he seemed satisfied as before,
and in time ascended the stupendous hill.
On his reaching the top his delight was
visible in a most eminent degree: he caressed his keeper and threw the dirt about
in a most playful manner.

"'Another elephant—a much younger
animal—was now to follow. He had
watched the ascent of the other with the
most intense interest, making motions all
the while as though he were assisting him by
shouldering him up the acclivity. When he
saw his comrade up, he evinced his pleasure
by giving a salute something like the sound
of a trumpet. When called upon to take

his turn, however, he seemed much alarmed, and would not act at all without force. When he was two steps up, he slipped, but recovered himself by digging his toes into the earth. With the exception of this little accident he ascended exceedingly well.

"'When this elephant was near the top, the other who had already performed his task extended his trunk to the assistance of his brother in distress, round which the young animal entwined his, and thus reached the summit of the hill in safety. Having both accomplished their task, their greeting was as cordial as if they had been long separated from each other, and had just escaped from some perilous achievement. They embraced each other and stood face to face for a considerable time, as if whispering congratulations. Their driver then made them saläm to the general, who ordered them five rupees each for sweetmeats. On this reward of their merit being ordered, they immediately returned thanks by another saläm.'"

The little audience were very enthusiastic

over these "delightful elephants" who acted so much like human beings, but of course they wished to know the meaning of "saläm" and "rupee."

"A saläm," said their governess, "is a very low and prolonged bow, and a rupee is Indian money valued at about forty cents of our coin.—Yes, elephants are passionately fond of sweet things, and two dollars' worth of candy for such a gigantic child would not equal more than a mere bite for one of you. It is astonishing, too, how daintily and carefully they will eat it. An elephant on exhibition was presented by one of his visitors with a package of candy wrapped in white paper, and he had no idea of putting it into his mouth just as it was. 'He curled up the end of his trunk and laid the package in the hollow of the curve; then he rubbed it with his finger until the paper was broken and the candy fell out on his trunk. He threw the paper away, gathered up the candy with his finger, and carried it to his mouth without dropping a single piece.'"

It seemed as though wonders about ele-

phants would never cease, and the children listened entranced.

"These animals," continued Miss Harson, "have been employed in India from
the earliest times in every possible way in
which strength and intelligence could be
made useful. On receiving an order they
will execute it without having any one to
watch them, and two elephants have been
seen battering down a wall at their keepers' request, with the promise of a reward
in the shape of fruit and brandy. A gentleman who watched them says that 'they
combined their efforts, and, doubling up
their trunks—which were guarded from injury by leather—thrust against the strongest part of the wall, and by reiterated
shocks continued their attacks, still observ-
ing and following with their eyes the effect of the thrusts; then, at last, making
one grand effort, they suddenly drew back
together, that they might not be wounded
by the ruins.' An order like this is always
given with the promise of some reward,
and trained elephants, judging from their
actions, readily understand the meaning of

various words to which they become accustomed. They will always answer to their names; and when a particular one is called, he makes a shrill noise, as much as to say, 'I am coming,' and goes at once to his keeper. Elephants will do almost anything for brandy or sweetmeats; but if, when the work is finished, there is any attempt to keep back the pay, the animal becomes furious."

"But, Miss Harson," said Clara, "how can an elephant understand that the things are given to him because he does the work?"

"It is managed," replied her governess, "by first showing the animal what he is to have, then urging him to the work, and as soon as it is done giving him the thing promised. He thus learns to connect his extra efforts with particular rewards, and is more ready to undertake a heavy piece of work. But in some cases neither reward nor punishment will avail, and Bishop Heber mentions a large elephant that was brought up to get on his feet again a poor old starved elephant that had fallen down.

'I was much struck,' he says, 'with the almost human expression of surprise, alarm and perplexity in his countenance when he approached his fallen companion. They fastened a chain round his neck and about the body of the sick beast, and urged him in all ways, by encouragement and blows, to drag him up, even thrusting spears into his flanks. He pulled stoutly for a minute, but on the first groan his companion gave he stopped short, turned fiercely around with a loud roar, and with his trunk and fore feet began to attempt to loosen the chain from his neck.' "

" I suppose," said Malcolm, " that he felt so sorry for the poor fellow."

" Yes, and indignant that the unfeeling owners should be willing to increase his sufferings. The elephant has also shown the same care and sympathy for human beings, and there is a story told of one that was marching with part of an English army in India just behind a heavy vehicle from which a soldier tumbled in such a way that the hind wheel would have gone over him in a second or two. But he was saved by the strength and ready wit of the elephant, who entirely of his own accord lifted up the wheel with his trunk and held it until the carriage had passed entirely clear of the man."

"I hope he had ever so much sweet-meats," cried Edith. "Don't you think he deserved it, Miss Harson?"

"He did indeed, and so did the nabob's elephant that carried his master, attended by numerous slaves, along the road from the palace, where many poor natives lay sick or dying of an epidemic disease. The slaves did not seem to care for their suffering fellow-creatures and would have gone right over them, but the more tender-hearted animal took the trouble to lift a number of them out of the way with his trunk, and stepped so carefully over the others that none were hurt at all."

"I just hope that old nabob caught the epidemic and died," exclaimed Malcolm, savagely. "He ought to."

"The story does not say that he did," was the quiet reply. "Besides, Malcolm, God does not often punish in this way.

His punishments, like his rewards, are generally a long time in coming—that is, as we
count time—and for this reason foolish people will think that they are not coming at
all. I am quite sure that the wicked nabob

ROYAL ELEPHANT WITH TRAPPINGS.

was punished, but it is not necessary that
we should know how or when.—In ancient
times," continued the young lady, "the
princes and rulers of India used to ride
in great state, with their families, upon ele-

phants that fairly shone with jeweled trappings; and one of these sovereigns rode on
an elephant through the streets of his capital, ' followed by twenty royal elephants for
his own ascending, so rich that in precious
stones and furniture they braved the sun.'
But now ' the stately animal is generally used
for the conveyance of the manifold servants
that wait upon the rich in India, or he is
laden with tents and tent-poles, or with water-bottles and pots and saucepans, and
all other paraphernalia of the kitchen,
slung about his body in all directions. His
appearance then is somewhat more ludicrous than dignified.' "

"Miss Harson," said Clara, suddenly,
"here is a very strange picture of a queer-
looking animal climbing upon an elephant
who has his trunk 'way up above his head,
and there are men on his back, and one
of them is poking something at the animal.
There is another elephant near by, and the
man on that one looks awfully frightened."

"Well he may," was the reply, "for the
' queer-looking animal' is that most ferocious of beasts a full-grown tiger, and he
22

has just sprung on the elephant in front.
The picture is a rough woodcut that makes
the tiger very queer-looking indeed."

"Do tigers ride on elephants too?"
asked Edith, with such a surprised face
that no one could help laughing.

"No, dear," said Miss Harson; "they
only spring on them as this one has done,
and try to tear them to pieces with their
terrible teeth and claws. But the elephant
is a match even for the tiger, and for that
reason he is used in India to hunt the
savage beast. You see how tall and thick
the great reeds are in the jungle where
the tiger makes his lair, and it requires the
elephant's strength to get through it, while
his keen scent discovers the prey before
any one has seen it, and his sagacity and
great height keep the hunter out of danger.
If the elephant were of no other use, he
would be invaluable as a means of ridding
the country of those fearful pests, which
make a point of devouring the inhabitants
whenever they can get at them. Not all
elephants are very courageous with tigers,
and a panic sometimes occurs as soon as

the detested animal appears. 'Occasionally
the hunter, with his rifle, is mounted upon
an elephant's back. The presence of the
tiger is generally made known by the elephants, which, scenting their enemy, become agitated and make that peculiar
trumpeting which indicates their alarm.
If the tiger moves, many of the elephants
become ungovernable; their trunks are
thrown up into the air; if they consent
to go forward, their cautious steps evince
their apprehensions. Those that remain
steady in such circumstances are considered
particularly valuable. If the motion of an
animal through the jungle is perceived, the
nearest elephant is halted, and the rider
fires in the direction of the waving rushes.
The tiger is sometimes wounded by these
random shots, and he then generally bounds
through the cover to the nearest elephant.
Very few elephants can then resist the
impulse of their fears. If the trunk—
which the animal invariably throws up as
far as possible out of reach—should be
scratched by the tiger, all command is lost.'
An elephant has sometimes been known to

catch a tiger on his tusks just as the animal
had made a spring, but this is not common.
If the tiger happens to fall, its ponderous
antagonist will instantly kneel upon its
body and fasten it to the ground with his
tusks. Elephants are trained to do this
with stuffed tigers, on which they are taught
to trample, and they never seem to forget
anything that they have once learned."

"Well," said Malcolm, "I do think that
elephants are the most wonderful animals
that ever lived, and I wish I owned one."

"You would have a white one, of course,
like the king of Siam?" said his governess,
laughing.

"Oh, Miss Harson," exclaimed Clara,
"will you not tell us something about white
elephants?"

"There is not much to tell, dear," was
the reply, "as there is no such thing, as I
have told you, as a really white elephant.
Those that are so called have a pinkish
tinge which is produced by disease, although
it is declared by an old writer that 'when
the king goes to court he has a train of
two hundred elephants, among which one is

white.' To this monarch's title—'king of the white elephant'—was added, 'Which elephant is the king of elephants, before whom many thousands of other elephants must bow and fall upon their knees.'"

"Did they do it?" asked Malcolm.

"If they did," replied his governess, "it was in the same way that the elephants belonging to another Eastern king made their obeisance. These animals were daily paraded in twelve companies splendidly adorned, 'the first elephant having all the plates on his head and breast set with rubies and emeralds, being a beast of wonderful stature and beauty. They all bowed down before the king.' This is explained by saying that on approaching the throne each driver pricked his elephant with a sharp instrument, and spoke to him at the same time, until he bent on one knee."

"Pooh!" said Edith; "that wasn't very smart, then, for an elephant. But it must have been nice," she added, "to see that great procession."

"It was certainly a brilliant sight," con-

tinued Miss Harson, "and elephants are
said to enjoy these parades, and to step in
a more stately manner when they are richly
adorned. An English traveler who saw
the king of Siam's elephants mentions
four white ones and gives the following
account of them: 'Within the first gate of
the palace is a very large court, on both
sides of which are the houses for the king's
elephants, which are wonderfully large and
handsome, and are trained for war and for
the king's service. Among the rest, he
has four white elephants, which are so
great a rarity, no other king having any
but he; and were any other king to have
one, he would send for it, and if refused
would go to war for it, and would rather
lose a great part of his kingdom than not
have the elephant. When any white elephant is brought to the king, all the merchants in the city are commanded to go
and visit him, on which occasion each individual makes a present of half a ducat—
which amounts to a good round sum, as
there are a great many merchants—after
which present you may go and see them at

your pleasure, although they stand in the king's house. Great honor and service are done to those white elephants, every one of them having a house with gold, and getting their food in vessels of gilt silver. Every day, when they go to the river to wash, each one goes under a canopy of cloth of gold or silk carried by six or eight men, and eight or ten men go before each, playing on drums, *shawms* and other instruments. When each has washed and is come out of the river, he has a gentleman to wash his feet in a silver basin; which office is appointed by the king. There is no such account of the *black* elephants, be they never so great, and some of them are wonderfully large and handsome.' It seems that the sovereign of Birmah had a white elephant too, but this one was said to be cream-colored. An unusual grunt from this important animal would take off the attention of the government from the most pressing affairs; for who could tell what might be working in the white elephant's mind and how far he might be able to see into the future? So these superstitious people spent their time in trying to explain what did not need any explanation, and left their business to take care of itself."

Then Miss Harson read to the children from a book about Siam * which she had been holding in her hand a description of a reception of a white elephant at the court of Siam :

" 'A few years ago two Siamese peasants of the up-country, far to the north, were ordered by the governor of the province to go out into the jungle and hunt for a white elephant. The astrologers having prophesied that the present reign would be especially lucky, and that several of these spotted elephants would be caught, constant vigilance had been enjoined on all the provincial officials of these regions, and a large bounty was promised to the finders of such a prize.

" 'Accordingly, leaving their homes and families, these poor men went out to live in the malarious jungle, wandering hither and thither for many weary weeks in vain, by day forcing their way through the rank un-

* Siam and Laos.

dergrowth, anxiously following the tracks of
the wild elephants up and down the streams,
living on the fruits of the trees and the fish
in the mountain-lakes, at night bivouack-
ing under the stars, each in turn watching,
while the other slept, to keep up the great
fire built to protect their resting-spot from
the fierce animals prowling about under
cover of the darkness. Thus day after day
and week after week they sought for the
coveted white elephant which should ensure to those who found him the richest
reward. At length, on the very point of
giving up in despair, they had turned their
faces homeward, when, all of a sudden, a
small beautifully-formed elephant was seen
at a distance, drinking. He was all muddy
and dirty, and at first sight appeared darker
than the ordinary color of this animal. But
some peculiarity in the skin aroused hope.
One of them said, "We will take him home
and give him a wash." This was done, and
to their great joy the whole body proved to
be of a pale Bath-brick color, with a few
real white hairs on the back. Indeed, competent experts pronounced it to be the

"fairest" ever caught within living memory.

" ' The whole kingdom was thrown into a state of the wildest excitement as the news spread. A fleet messenger bore the official document with the formal announcement down the river to Bangkok. The king loaded his ears with gold. Each person in any way connected with this great capture received some token of royal favor. The poor finders were loaded with honors and emoluments, at one step taking their places among the nobles of the kingdom and receiving royal gifts and grants of land.

" 'A day was fixed for the reception of the royal stranger at the capital, and His Majesty with his entire royal retinue went up the river sixty miles, some days in advance, to meet the illustrious captive.

" ' Very early in the day the whole city was astir. The most intense excitement prevailed. It was a great fête occasion. Old and young thronged the verandas of the houses. Crowds of country-folks from miles around flocked to the river. Near the palace-grounds, as the time drew near for

SIAMESE WORSHIPING THE WHITE ELEPHANT.

the procession to approach, there was much
running to and fro, officials on horseback
galloping about, soldiers and marines in
uniform. The national air, played by a
brass band, heralded the approach of "the
conquering hero." A temporary stable had
been erected for this illustrious captive just
outside the palace-grounds. He was mounted on a platform, and his hind leg was attached by a rope to a white post. Here,
after numerous washings by pouring over
him tamarind-water to cleanse away all possible impurities, the new elephant was pub-
licly baptized and received official title as a
grandee of Siam. He was then brought into
the palace-precincts and assigned a royal
stable and numerous attendants, who serve
him with the respect shown to royalty itself,
and who generally approach on their hands
and knees to feed and groom him. This
elephant, however, is young, lively and good-
natured, and makes saläms by raising his
trunk straight and high above his head to
all well-dressed visitors in a way which
quite scandalizes his keepers, who have
taught all the other royal elephants to re-

serve that salute solely for the king. Were
he not too royal to be whipped, this merry
grandee might soon be taught to recognize
the honor due to royalists.

"'In time past these beasts were wor-
shiped by king and people; their stables
were palaces; they were fed from golden
dishes and wore heavy gold rings upon
their tusks, and were fettered with golden
chains. Yet even now the populace fall
with their heads to the ground as these
white elephants are led out richly capar-
isoned on state occasions, while the royal
officers, and even the king himself, always
make them obeisance in passing.'"

CHAPTER XVII

"I SUPPOSE," said Clara, when the children were gathered again in their favorite place conveniently near Miss Harson, "that there isn't any more about elephants except what you were going to tell us about some menagerie elephants?"

"I think," replied her governess, with a smile, "that we are almost as well acquainted with these large and useful animals as we can hope to be—with the exception, perhaps, of some few particular ones that have been on exhibition. It is wonderful to think of their being carried around the country in cars like human beings and made to do a variety of things that are quite foreign to their usual habits."

"Do they take elephants on cars, Miss Harson?" asked Edith, as though she ex-

350

pected to see two or three of them seated in the next train she entered.

"Yes, Edie, but not as they take us. The poor creatures are very much cramped in what is called an 'elephant-car,' five of them being stowed in at once; and they usually travel at night, so that no part of the day may be wasted. All this must seem very strange and alarming to the unwieldy animals, and with their great caution and unwillingness to try anything new it is a wonder that they can be taken about from one place to another with so few mishaps."

"The people who own the elephants and things," said Malcolm, "must make lots of money, because every one pays to go and look at 'em."

"They take in a great deal of money," replied Miss Harson, "but it is not all gain. It costs a great deal to feed large animals and to take proper care of them. Men have to be hired at very high wages as keepers or attendants, for there are very few who understand this business and can get control over the animals. I suppose you have not happened to think of the labor of washing and dressing them?"

Malcolm could not say that he had, and his sisters were very much surprised to hear that animals were washed and dressed.

"I do not mean," continued the young lady, " that they are washed and dressed as we are, but many of them, like elephants, camels and giraffes, have showy trappings put on them, and, although there are no bath-tubs for their special use, they are regularly scrubbed off. I remember seeing a picture some time ago in which a large elephant was having a bath. He looked perfectly contented as he squatted on the ground close by a hydrant while one man played the hose over him and another, kneeling on his back, washed him off with a long broom. The enormous animal seemed pleased with these attentions, and the men saved him the trouble of squirting himself over with water from his trunk. Then, too, the cutting of his nails is quite an undertaking."

"Oh, Miss Harson!" said Clara, who

thought her governess must mean this for a joke; "do elephants really have their nails cut?"

"I will read you what I found about it in a book," was the reply; "and I do not wonder at your surprise, for I was very much surprised myself. It seems that 'three times a year, at least, each one of these monsters must have his hoofs cut and trimmed into good shape—once in the spring, once when traveling with the circus in summer, and once more after the huge beast has returned to winter quarters. The sole of the elephant's foot becomes gradually covered during the year with a thick substance resembling horn, much like the three great toe-nails. This, if allowed to grow too dense, is apt to crack and make the beast lame. Accordingly, one of the keepers stations the elephant in the ring and bids him balance himself on three legs, while he stretches out the other behind him, resting it on a tub or a box. With a carpenter's large drawing-knife the hoof is then attacked and quickly shaved down. Sometimes pieces of the bony

23

substance five or six inches long and nearly as thick are cut off without the elephant's feeling any pain whatever or the knife taking too much from the sole. Frequently pieces of glass, nails, splinters, and the like, are found imbedded in the growth, and these it is very important to have extracted, lest they should work their way upward and fester in the foot. Recently a nail was discovered and pulled out from the foot of the elephant Pallas that only came to light after three inches of the hoof had been cut off. When the first rough going over is completed, the keeper, with a smaller knife, trims each nail into handsome shape—its cleanness and new color quite improving the animal's appearance—covers any small wounds with tar and dismisses the patient. It takes six hours to do this curious job in a proper manner, and the keeper is quite tired out when two beasts have received his attentions.' "

"Here is a pleasant story," continued Miss Harson, "about an elephant belonging to a menagerie which outwitted the

people of the village. The story is called
'Regulating the Elephant:'

"'Everybody had heard that the great
elephant was loose, and several families
whose gardens he had torn up and whose
boys he had trampled upon were certain of
it. There was a great excitement, and the
town held a meeting to decide what should
be done. They did not want to exterminate him : in fact, many of them did not
believe they could exterminate him, for he
was a pretty big elephant. Besides, he was
useful in his proper place—in shows, in India and in story-books.

"'"Our best plan is to try and regulate
him," said an enthusiastic speaker. "Let us
build toll-gates all along the route we find
he is going to take, and make him pay—"

"'"Yes, but that leaves him roaming
around," said an old woman, "and I don't
want my boy killed."

"'" Keep your boy away from him ; that's
your business. Why, madam, don't you
know that an elephant's hide and tusks are
valuable for mechanical purposes, and that
he is useful in India? Besides, there's the

toll he will pay. We shall by this means
get money into the public treasury to build

REGULATING THE ELEPHANT.

schools for a good many boys who are not
trampled to death."

"'"That's the plan! Regulate him! regulate him!" shouted the crowd.

"'So they appointed a great many committees, and drafted constitutions and by-laws, and circulated petitions, and by the time the elephant killed several more boys and trampled down a quantity of gardens they had erected very comfortable toll-houses for the gatekeepers and gates for the elephant; and they waited in great satisfaction to see the animal regulated.

"'Slowly the great feet tramped onward; slowly the great proboscis appeared in view; and, with a sniff of contempt, the elephant lifted the gate from its hinges and walked off with it, while the crowd stared after him in dismay.

"'"Well!" exclaimed the keeper, catching his breath; "we haven't made much money so far, but the regulating plan would have been first rate if the elephant had not been a little stronger than the obstruction."'"

This was certainly something new about elephants, and the little audience had heard so many marvelous things that they would scarcely have been surprised at anything now.

"Then, you know," said Miss Harson, "they must have their lessons."

"Out of a book?" asked Edith, in a state of great bewilderment.

"They have not got quite so far as that yet," was the laughing reply, "but they are taught to stand on their heads and to turn somersaults, ring a bell with their trunks, play at see-saw, and do a variety of other things. The funniest part of it is that they have been seen practicing these antics by themselves when they supposed that no one was looking at them. It is astonishing how much an elephant can be taught to do and how well he remembers what he is taught. These performing elephants, as they are called, always draw large crowds of people to see them, and they are very valuable to their owners. Several years ago there was in this country a large performing elephant named 'Queen' who was not always very good-tempered, and in a sulky fit she would refuse to play at all. But two of the circus-people owned a dear little baby-boy, who was taken one day to see the big elephant that was not very

pleasant, and he immediately put out his
little arms and cooed at it. Queen was
surprised for a moment, but presently there
was a strange rumbling sound that meant
pleasure, and the big elephant followed the
boy with her eyes as he was carried away.
Before long Baby had his chubby arms
around Queen's trunk, and this delighted
her very much. He would stick his finger
in her eyes and take all sorts of liberties
with the great creature, but she never resented it, and seemed perfectly happy when
the little fellow was with her. When he
got old enough to toddle about, Donald
would make the big elephant lie down by
pushing at her, when she would tumble on
one side to please him. Then he climbed up
on her head in high glee, letting his little
fat legs hang down in his heavy friend's
mouth. By and by they got to playing-
blocks, and Queen would take up one in
her trunk and lay it on the others, and then
wait for Donald to lay his. When the pile
was finished—perhaps a castle and a fence
—the little boy would laugh mischievously and knock it down. Then Queen rum-

bled with delight, and this seemed to be
the best part of the play. It was a constant enjoyment to the circus-people of all
kinds to watch these strange playmates,
neither of whom seemed to notice any
one that was looking on.

"You were surprised," continued the
young lady, "when I told you of the elephant that took care of the baby in India,
but no wee boy ever had a more devoted
attendant than this great, lumbering, affectionate Queen. The two would lie down
together and go to sleep in the most perfect security, and it was a very common
thing to see the elephant standing with the
mite of a child curled up in her trunk, on
which he had climbed. Both seemed to
have the happiest possible times, and little
Donald's laugh always appeared to delight
his huge friend. But after a few days' ill-
ness the little boy died, and, from being
uneasy at first on missing him as day after
day passed without his appearing, the great
animal became frantic with grief. She
treated her keeper, to whom she had been
attached, as though she suspected him of

having made way with her pet, and she refused for some time to take her part in the
performances. She gave a great deal of
trouble to her keeper and owners, because
the other performing elephants, who acknowledged her as their sovereign, were
disposed to avenge her fancied wrongs,
and even after she was prevailed upon to
take up her duties again she was a sulky,
disagreeable animal."

"But wasn't it nice of her," said Edith,
who was almost crying about Donald, "to
be sorry for the dear boy?"

"Yes, dear, it showed a very affectionate
nature and a strong memory, neither of
which we seem to expect in an elephant.
But I think we generally expect too little
from animals.—As to Queen, after a while
there came to the menagerie another baby,
which stood on four legs and had a funny
little trunk and did not seem to know what
to do with itself. It was Queen's own
baby, but she did not appear to notice it
until the other elephants went frantic with
delight, and stood on their heads, and performed all the queer capers they had been

taught. The bewildered baby stood perfectly still during this strange welcome,
when suddenly its mother caught sight
of it, and with a shrill yell of delight, she
snatched it up in her trunk and flung it
as far as she could."

"How horrid!" exclaimed Clara. "Did
she kill the poor little thing?"

"No; 'the poor little thing'—who was
not much less than three feet high—probably thought it had got into a queer sort
of world, but it made no remarks and just
stayed where its mother had thrown it.
Queen was a changed elephant after that;
she took the greatest possible comfort in
her baby and became mild and obedient.
The other elephants seemed to form themselves into a band of nurses, and it was
very amusing to see the care they took of
that precious baby. They would not allow
it to cross a bridge until they had tried it to
see if it was quite safe, and then they
would caress it with their trunks to en-
courage it. If ever there was a spoiled
elephant-child, that was certainly one;
and besides the adulation it received in the

family circle, it had hundreds of visitors
every day, and was stuffed with cake and
candy and fruit and peanuts, until the
wonder was that it managed to live. It
was the first elephant born in captivity—
the only baby-elephant ever seen in this
country—and people could not make
enough of it."

"Couldn't we see it, Miss Harson?"
asked the children, excitedly.

"Not that baby-elephant, because it is
quite grown up now; but if there is
another one, I shall hope to take you.—
Another wonderful menagerie elephant
was the huge African Jumbo, who after
spending about three years in this country
was killed by a railroad-train some time
ago. He was the largest elephant ever
brought here, and it was a very hard and
expensive piece of work to bring him—
that is, to bring him from England; for
Jumbo was caught by some Arabs when he
was a baby, and he had been at the Zoolog-
ical Gardens in London for a great many
years, so that he knew nothing at all of
forest-life in Africa. Jumbo was very

gentle and a great favorite with the English children, who loved to ride on his back
and feed him with buns and candy and
peanuts ; but after a while he had fits of
anger, and was sold to Mr. Barnum for a
large sum of money.   It was thought that
he would do better under different treatment, and this proved to be the case, for in
America he was always gentle and affectionate.   The children went wild over him
and delighted in riding on his back, while
he and his keeper were devoted friends.
They had been together ever since Jumbo
was three years old ; and when he died, he
was about twenty-five.   Such a valuable elephant had every possible care and wore rich
and beautiful trappings, but he was not allowed to parade the streets like more common animals.   He had his own special car
for traveling, and his daily rations were
two hundred pounds of hay, two bushels of
oats, a barrel of potatoes, ten or fifteen
large loaves of bread and two or three
quarts of onions, besides the nuts and
sweet things with which he was constantly
stuffed by visitors.   But he was always in

perfect health, and seemed to be a happy and contented elephant."

"And was he killed, Miss Harson?" asked Edith, sorrowfully.

"Yes, dear; he was killed on a railroad-track, where he and a companion-elephant had unfortunately gone to walk just after delighting people with their amusing performances. Hearing a loud, strange noise, the people who were preparing to pack up the show rushed to the car-track near by, and there was Jumbo crushed and dying, another elephant with a broken leg, and part of a railroad-train thrown off the rails into a ditch. It was a terrible scene, and after that one wild cry Jumbo was perfectly quiet. His keeper wept on his dead body as if he had lost a human friend, and he declared that if Jumbo was only an elephant he was, as the saying is, every inch a king."

There was some weeping done at the end of this story, and Miss Harson comforted her little charges by telling them that Jumbo had been stuffed and could be seen in that way, and that on the first opportunity they should have a sight of this world-renowned elephant.

But the children were not at all sure that they would not feel worse—"because he was not alive, you know"—and the talk about elephants ended with their deeply considering the matter.

THE END.

www.ingramcontent.com/pod-product-compliance
Lightning Source LLC
Chambersburg PA
CBHW021401210326
41599CB00011B/959